Lecture Notes in Physics

T0238828

The Lecture Notes in Physics

The series Lecture Notes in Physics (LNP), founded in 1969, reports new developments in physics research and teaching – quickly and informally, but with a high quality and the explicit aim to summarize and communicate current knowledge in an accessible way. Books published in this series are conceived as bridging material between advanced graduate textbooks and the forefront of research and to serve three purposes:

- to be a compact and modern up-to-date source of reference on a well-defined topic

- to serve as an accessible introduction to the field to postgraduate students and nonspecialist researchers from related areas

- to be a source of advanced teaching material for specialized seminars, courses and schools

Both monographs and multi-author volumes will be considered for publication. Edited volumes should, however, consist of a very limited number of contributions only. Proceedings will not be considered for LNP.

Volumes published in LNP are disseminated both in print and in electronic formats, the electronic archive being available at springerlink.com. The series content is indexed, abstracted and referenced by many abstracting and information services, bibliographic networks, subscription agencies, library networks, and consortia.

Proposals should be sent to a member of the Editorial Board, or directly to the managing editor at Springer:

Christian Caron
Springer Heidelberg
Physics Editorial Department I
Tiergartenstrasse 17
69121 Heidelberg / Germany
christian.caron@springer.com

K.-L. Klein
A. L. MacKinnon (Eds.)

The High Energy Solar Corona: Waves, Eruptions, Particles

 Springer

Editors

Karl-Ludwig Klein
Observatoire de Paris
Sect. de Meudon
5 place Jules Janssen
92195 Meudon CX, France
ludwig.klein@obspm.fr

Alexander L. MacKinnon
University of Glasgow
Dept. Adult and Continuing Education
11 Eldon Street
St. Andrew's Building
Glasgow, UK G3 6NH
a.mackinnon@educ.gla.ac.uk

K.-L. Klein and A. L. MacKinnon (Eds.), *The High Energy Solar Corona: Waves, Eruptions, Particles*, Lect. Notes Phys. 725 (Springer, Berlin Heidelberg 2007), DOI 10.1007/ 978-3-540-71570-2

ISBN 978-3-642-09076-9   e-ISBN 978-3-540-71570-2

ISSN 0075-8450

Springer is a part of Springer Science+Business Media
springer.com
© Springer-Verlag Berlin Heidelberg 2007
Softcover reprint of the hardcover 1st edition 2007

Cover design: eStudio Calamar S.L., F. Steinen-Broo, Pau/Girona, Spain

Preface

The Community of European Solar Radio Astronomers (CESRA) organizes workshops on investigations of the solar atmosphere using radio and other observations. These workshops are intended to bring together different communities of observers and theoreticians in order to discuss current problems of the structure and dynamics of the solar atmosphere, its impact on, and relationship with, the heliosphere. For the 2004 workshop, we decided to give special emphasis to three topics, each of interest in its own right but also feeding into our overall picture of the Sun: small-scale energy release and fast particle acceleration and transport; large-scale disturbances, their origin and consequences; and radio pulsations as diagnostics of solar atmosphere plasma parameters. Radio observations offer a distinctive view of each of these phenomena, but each benefits also from a multi-wavelength perspective drawing particularly on findings from current space missions: RHESSI, TRACE, SoHO. We invited a group of speakers who together represent the multi-wavelength totality of these phenomena while concentrating on the role of radio wavelengths. They were encouraged to emphasize fundamentals and incorporate a tutorial element into their presentations, so that this book, growing out of the contributions to the Workshop, should have lasting value as well as discussing topics that generate excitement at the moment. In particular it highlights some of the areas of outstanding interest in solar radio astronomy in the run-up to the FASR facility.

The CESRA workshop 2004 was organised by a scientific committee composed of H. Aurass and G. Mann (Germany), M. Karlický (Czech Republic), K.-L. Klein (France), A. MacKinnon (United Kingdom), and A. Stepanov (Russia), in close cooperation with the Solar Physics Section of the European Physical Society (EPS) and the European Astronomical Society (EAS). The local organisers were A. MacKinnon, L. Bone, L. Fletcher, R. Galloway, J. Khan, E. Kontar, P. Mallik (University of Glasgow), and P. Wood (University of Saint Andrews). The workshop was held at Sabhal Mor Ostaig (Isle of Skye, Scotland). Carlotta Graham and the rest of the staff there worked hard to foster a very pleasant atmosphere that will long be remembered by the participants. This meeting would not have been possible without the financial support of the *Royal Astronomical Society*, the *Highland Council*, the *British Council*, and the *Observatoire de Paris*. The Highland Council in particular provided support aimed at simplifying the journey to an outstanding, if slightly remote, location. Glasgow University's Departments of Physics and Astronomy and of Adult and Continuing Education provided various sorts of logistical support. The editors are indebted to the referees of the papers in this book. Among them were M. Aschwanden, G. Aulanier, I. Cairns, C. Chiuderi, A.G. Emslie, B. Roberts, and M. Velli.

Meudon and Glasgow,   *Karl-Ludwig Klein & Alexander MacKinnon*
February 2007   *on behalf of the Scientific Organising Committee*

Contents

Introduction: The High-energy Corona – Waves, Eruptions, Particles
Karl-Ludwig Klein and Alexander MacKinnon 1
1   Particle Acceleration During Flares 1
2   Large-scale Disturbances 5
3   Waves in Coronal Magnetic Field Structures
     and Plasma Diagnostics 9
4   Outlook 10
References 11

Part I Particle Acceleration During Flares

Magnetic Complexity, Fragmentation, Particle Acceleration and Radio Emission from the Sun
Loukas Vlahos 15
1   Introduction 15
2   Classical Models for Energy Release and Particle Acceleration 17
3   Energy Release and Particle Acceleration
     in Complex Magnetic Topologies 23
4   Radio Emission from Simple
     and Complex Magnetic Topologies 26
5   Summary 29
References 30

Review of Selected RHESSI Solar Results
Brian R. Dennis, Hugh S. Hudson, and Säm Krucker 33
1   Introduction 33
2   Instrumentation 34
3   Soft X-rays 35
4   Hard X-rays 43
5   Gamma Rays 51

6   Flare/CME Energetics 58
7   Conclusions... 60
References 62

RHESSI Results – Time for a Rethink?
J. C. Brown, E. P. Kontar, and A. M. Veronig 65
1   Introduction 65
2   Imaging Discoveries and Issues 66
3   Temporal Domain Discoveries and Issues 71
4   Spectral Discoveries and Issues 74
5   Conclusions... 79
References 79

**Small Scale Energy Release
and the Acceleration and Transport
of Energetic Particles**
Hugh Hudson and Nicole Vilmer 81
1   Introduction 81
2   Some New Observational Constraints 83
3   Acceleration and Propagation Mechanisms....................... 94
4   Radio Emission Mechanisms 97
5   Conclusions... 98
Appendix 99
References 99

Part II Large-scale Disturbances

**Large-scale Waves and Shocks
in the Solar Corona**
Alexander Warmuth 107
1   Introduction 107
2   The Physics of MHD Waves and Shocks 109
3   Signatures of Coronal Waves 112
4   Association with Type II Radio Bursts 119
5   The Physical Nature of Coronal Waves 120
6   Causes of Coronal Waves 127
7   Relevance of Coronal Waves to Other Areas
     of Solar Physics 130
8   Conclusions...132
References 134

**Energetic Particles Related with Coronal
and Interplanetary Shocks**
N. Gopalswamy 139
1   Introduction 139
2   Type II Radio Bursts and Shocks...............................141

3   Type II Bursts and SEP Events 146
4   CME Interaction 148
5   CME Interaction and SEPs: Case Studies 149
6   SEP Intensity Variation: Statistical Study 152
7   SEP Intensity and Active Region Area 154
8   Discussion and Conclusions 156
References 158

Particle Acceleration at the Earth's Bow Shock
David Burgess 161
1   Introduction 161
2   Electron Acceleration at the Quasi-perpendicular
    Bow Shock 164
3   Quasi-parallel Shocks and Acceleration 176
4   Transients at Shocks: Hot Flow Anomalies..................... 185
5   Summary... 188
References 188

**On the Existence of Non-maxwellian Velocity Distribution
Functions in the Corona and their Consequences
for the Solar Wind Acceleration**
Milan Maksimovic 191
1   Introduction 191
2   The Principal Features of Observed Solar Wind Electron
    Distribution Functions 192
3   Modelling the Consequence of Non-thermal Distributions
    in the Corona 196
4   Concluding Remarks... 201
References 201

**Recent Research: Large-scale Disturbances, their Origin
and Consequences**
Gottfried Mann and Bojan Vršnak 203
1   Introduction 203
2   Topics 204
3   Take-off and Propagation of CMEs 204
4   Origin and Propagation of Shocks 208
5   The Role of Flares, CMEs, and Shocks
    in Particle Acceleration 209
6   Coronal and IP Plasma Diagnostics Offered
    by the Radio Emission 212
7   Conclusion 214
Appendix: The List of Participants 215
References 216

Contents

Part III Plasma of the Solar Corona

Quasi-periodic Pulsations as a Diagnostic Tool for Coronal Plasma Parameters

V. M. Nakariakov and A. V. Stepanov 221

1 Introduction 221
2 MHD Modes of a Plasma Cylinder 223
3 Standing Longitudinal Waves 231
4 Ballooning Modes 233
5 Damping 234
6 Loop Plasma Diagnostics (with Examples) 236
7 Equivalent Electric Circuit 242
8 Conclusions 247
References 248

Pulsating Solar Radio Emission

Alexander Nindos and Henry Aurass 251

1 Introduction 251
2 A Brief Overview of Radio Pulsations 253
3 Microwave pulsations 257
4 Decimetric Pulsating Emission 261
5 Metric Pulsating Emission 266
6 Summary and Future Work 273
References 275

Index 279

Introduction: The High-energy Corona – Waves, Eruptions, Particles

Karl-Ludwig Klein[1] and Alexander MacKinnon[2]

[1] Observatoire de Paris, LESIA-CNRS UMR 8109, 92195 Meudon, France
ludwig.klein@obspm.fr
[2] DACE/Physics & Astronomy, University of Glasgow, Glasgow G3 6NH,
Scotland, United Kingdom
a.mackinnon@educ.gla.ac.uk

Flares and coronal mass ejections (CME) are the most violent manifestations of solar activity. They are the consequence of the explosive conversion of energy stored in coronal magnetic fields into plasma heating, the kinetic energy of supra-thermal to high energy particles and the mechanical energy of magnetic structures that are propelled through the corona and into interplanetary space. The corona is the seat of activity on a large variety of spatial and temporal scales, and presents us with a unique opportunity to sound the relevant plasma phenomena with imaging and spectrographic observations, as well as with in situ measurements near the Earth. This book addresses three key features: eruptions of magnetic structures, associated large-scale perturbations such as propagating waves and shocks, and energetic particles which carry a large part of the energy released during flare/CME events.

1 Particle Acceleration During Flares

The transient brightenings indicating the occurrence of a solar flare occur across the electromagnetic spectrum and involve a wide range of heights in the solar atmosphere. For many years the chromospheric Hα line was almost the only tool for their study; hence the old name, "chromospheric flare". Flare X-ray, EUV and radio emissions extend the study of flares into coronal regions where the flare originates, and also reveal the phenomena of particle acceleration that seem to be a central part of the flare process.

In order to introduce concepts, we show in Fig. 1 a widely used cartoon displaying key processes and the places and manners in which they are thought to occur (see the chapter by Hudson & Vilmer for further cartoons). Details are beyond doubt more involved, but this cartoon provides a useful framework for discussion. Energy stored in stressed coronal magnetic fields is thought to be explosively released in a current sheet, by magnetic reconnection of oppositely

K.-L. Klein and A. MacKinnon: *Introduction: The High-energy Corona – Waves, Eruptions, Particles*, Lect. Notes Phys. **725**, 1–11 (2007)
DOI 10.1007/978-3-540-71570-2_1

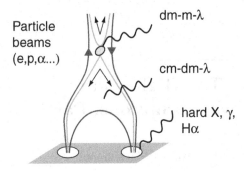

Fig. 1. Cartoon scenario of magnetic reconnection, energetic particles and radiative signatures during a flare

directed magnetic field lines. As indicated by arrows, fast particles may travel away from the acceleration region in both upward and downward directions. Upward moving particles escape to interplanetary space; those moving downwards towards the chromosphere along magnetic field lines may also become trapped in closed magnetic fields.

The particles interacting with the solar atmosphere can be remotely sensed through different types of electromagnetic radiation. Key to tracing the paths of electrons in the corona are the so-called "type III" radio bursts. They are produced by electron beams, which excite Langmuir waves in the ambient plasma through beam-plasma instabilities. These Langmuir waves are subsequently converted into electromagnetic radiation at the local electron plasma frequency or its harmonic. If the electron beams propagate outward towards interplanetary space, this emission consists of a short pulse whose frequency decreases as the beams proceed to regions of decreasing ambient electron density. When the electron beams propagate downward from the acceleration site, into a plasma of increasing electron density, the radio emission frequency increases in the course of time. In opposition to type III bursts, these features are called "reverse slope" bursts. In the solar corona type III bursts are mostly seen at metric and longer wavelengths. They can be tracked to Earth orbit at hectometric to kilometric wavelengths from space. Reverse slope bursts are observed in increasing numbers as the wavelengths become shorter (decimetre range). These kinds of radio emission therefore give an idea on the localisation of the acceleration region. But the coherent radiation process implies that the radiation spectrum carries little direct information on the energy content of the radiating electrons.

In the low corona the regions of closed magnetic field lines in Fig. 1 are characterised by magnetic field strengths >100 G (10^{-2} T). In such fields, mildly relativistic electrons (energies > 100 keV) radiate gyro-synchrotron emission at centimetric and shorter wavelengths. Those electrons which precipitate into the dense chromospheric footpoints are thermalised and simultaneously emit hard X-rays through electron-ion and, at relativistic energies, electron-electron bremsstrahlung. The most widely used models of hard X-ray emission, known as "thick target" models, assume that electrons stop in this way effectively instantaneously. X-ray emission can also be produced in more tenuous regions in the coronal loops. Similarly, protons and heavy ions accelerated in the corona to energies up to several tens of MeV/nucleon impinge onto the dense chromosphere and emit gamma-ray lines in the (2-8) MeV range when interacting with ambient nuclei. Relativistic protons create secondary particles, among them pions which decay into photons, electrons and positrons. Pion decay products contribute to continuum emission above a few tens of MeV.

Hα emission, as well as infra red and white light, come also from the footpoints of flaring loops. These emissions are secondary signatures of the energy release during flares, because they are excited by heat transport or energetic particles coming from the primary acceleration site. The footpoints of the 2D loops of Fig. 1 are expected to be bright in Hα, and to form elongated ribbons if a series of such loops is involved in flaring. The most clearly defined ribbons are observed in the aftermath of a filament eruption, where the coronal current sheet sketched in Fig. 1 is thought to be formed as the filament rises and leaves behind a region of low magnetic pressure.

Contributions in this volume address both theories of solar particle acceleration, and the observations that reveal its properties. Various acceleration processes have been discussed in the literature. Plasma inflow into the current sheet gives rise to a $V \times B$ convective electric field, maintained throughout the reconnection region by dissipative processes. This enables "DC" (i.e. not rapidly varying) electric field acceleration. Turbulence and shocks generated by magnetic reconnection, i.e. when jets flowing out of the reconnecting current sheet impinge onto the ambient plasma, are further mechanisms which are frequently discussed.

Loukas Vlahos discusses the physics of particle acceleration in solar magnetic field configurations, emphasising the role of fine structure that develops on scales far below those sketched in Fig. 1. He first presents and discusses popular flare models, but stresses the point that all relevant scales are far below those which are accessible to contemporary observations. In fact the convective motions which make magnetic flux emerge into the corona and shape coronal loops take place on small scales, and more or less randomly. In such cases multiple current sheets are expected in the corona, rather than a single one as depicted in Fig. 1. Such fine structure may also result as the energy release process spreads in the corona in the "avalanche" manner. Vlahos presents numerical simulations of such structures and of the behaviour of charged particles in such configurations.

He compares the predictions of these models with radio observations of solar flares. Novel radio instrumentation should be able to localise multiple small-scale sources especially at the start of flares. Coronal evolution with multiple small-scale energy release events could also explain long-duration (hours to days) radio emission that is observed in the aftermath of some flares or even without a conspicuous flare.

Radio waves, gamma rays and (hard) X-rays are the clearest diagnostic of energetic particles in the solar atmosphere. Hard X-ray (bremsstrahlung) and microwave (gyro-synchrotron) emission in particular give the most detailed diagnostics of the distribution function of flare accelerated electrons. The Ramaty High Energy Solar Spectroscopic Imager (RHESSI) spacecraft now provides for the first time images in the photon energy range from a few keV to several MeV, including thermal emission at soft X-rays, hard X-rays from energetic electrons and gamma-rays from energetic ions. RHESSI results are discussed in the chapters by Dennis et al., Brown et al., and Hudson & Vilmer.

A brief description of the instrument is given in the chapter by Brian Dennis, Hugh Hudson and Säm Krucker. Because of its high spectral resolution RHESSI allows for the most convincing separation of thermal and non thermal X-rays so far, with frequently observed thermal spectra from coronal sources, and non thermal spectra, at photon energies above typically 20-30 keV, from footpoints. This is consistent with the "thick target" model. The observation of a line complex from Fe and Ni ions near photon energies of 7 keV constitutes a new diagnostic of plasma temperature and elemental abundances during flares. Among new observations is the potential evidence for reconnection in coronal current sheets, similar to the scenario depicted in Fig. 1: RHESSI observed a pair of hard X-ray sources above the loop top, with one source moving upward, the other downward, and maximum temperature being measured in between. This corresponds to the scenario where freshly reconnected field lines retract upward and downward from the reconnection region, where temperature is indeed expected to be highest. RHESSI has also given new insight into gamma-ray spectroscopy. Dennis et al. summarise the principles of the analysis, and illustrate how RHESSI's spectral resolution e.g. of the 511 keV electron-positron annihilation line allows new studies of the dynamics of the low atmosphere during flares, thanks to the possibility to measure details of the line shape.

The greatest surprise of RHESSI concerned the locations of emissions of gamma-ray continuum sources, produced by mildly relativistic electrons, and nuclear line emission from protons. The close similarity of time profiles of hard X-ray and gamma-ray line emission had led to the expectation that both emissions come from the same sources. But RHESSI showed that the sources of bremsstrahlung gamma-rays from relativistic electrons and of 2.223 MeV line emission from neutron capture by ambient protons are different. The energetic neutrons result from nuclear interactions in the dense chromosphere. It seems that this intriguing observation holds a clue to the understanding of the particle acceleration process itself.

But surprises come not only from imaging. Traditional methods of spectral analysis may also be questioned by the new observations. These problems make John Brown, Eduard Kontar and Astrid Veronig wonder about the need to rethink well-known ideas on hard X-ray emission. Until recently, observed hard X-ray spectra were adequately represented by power laws in photon energy. RHESSI data reveal details apparently implying local minima in emitting electron energy distributions, irreconcilable with the thick target model. Pure thick target modelling of hard X-rays is also shown to be a problem when the spectra are compared with those of electrons measured directly in space. Another concern for our understanding of the emission are the huge amounts of electron fluxes that are required to account for the observed emission. This may be in conflict with the narrow traces of the presumed footpoints of flaring loops observed at UV wavelengths, if these emissions are explained by the excitation by electron beams guided by the freshly reconnected loop field lines. So while RHESSI-observed morphologies confirm the picture of particle acceleration during magnetic reconnection and of non thermal hard X-ray and gamma-ray emissions from particles guided along the reconnected loops into the dense chromosphere, the attempts at quantitative understanding reveal new questions on the radiation process itself.

Hugh Hudson and Nicole Vilmer discuss recent and ongoing research in the field of X-ray, gamma-ray and radio analyses of particle acceleration during flares. They give a qualitative overview of flare scenarios and indicate observational consequences, such as flare ribbons outlining the footpoints of an arcade of freshly reconnected magnetic loops. Basics of type III emissions and of gyrosynchrotron radiation are presented. Ongoing research work addresses radio diagnostic of time-extended particle acceleration (tens of minutes), spectrographic and imaging diagnostics (radio, X-rays) of energy release during flares and of particle trapping. The young and developing field of sub-millimetre observations is highlighted. It is of importance for the detection of relativistic electrons during flares, and of thermal emission from the low atmosphere. The first observations show some unexpected spectral features. Besides individual large flares, small flares are important because they are probably a much more frequent phenomenon which could be related to quasi-continuous energy release in the corona. New insight is brought by RHESSI on the locations and distribution of these events over the solar activity cycle. It is pointed out that the energy content of the microflares detected in the first months of RHESSI observations, i.e. in a period of high activity, is surprisingly high. This will be a major topic during the present phase of low solar activity.

2 Large-scale Disturbances

In relationship with flares and coronal mass ejections (CMEs), a variety of travelling disturbances is observed in the solar atmosphere. At radio wavelengths (dm and longer) type II emission is the longest known feature attributed to coronal shock waves. The emission consists of one or two narrow spectral bands that gradually drift towards lower frequencies. The drift rate is much slower than for type III bursts, and this is ascribed to an exciter that propagates at much lower speed than the type III emitting electron beams. An example of both burst types is shown in Fig. 2. The frequency drift can be translated into the speed of the exciter propagating through a plasma of decreasing electron density, by use of a density model inferred from independent observations. For type II bursts in the corona typical speeds are of the order of 1000 km/s, consistent with super-alfvenic MHD motion. The type II burst of Fig. 2 consists of two bands with a frequency ratio of about 2:1, which are interpreted as electromagnetic emission at the local electron plasma frequency and its harmonic.

Large-scale perturbations that propagate away from a flare were also observed in the Hα line since the 1960s and are referred to as "Moreton waves". They were interpreted as the trace in the chromosphere of three-dimensional shock waves whose coronal parts generate the type II burst radio emission. Later the EUV imager EIT aboard the SoHO spacecraft revealed the existence of large-scale travelling disturbances in this spectral range, too. They are termed "EIT waves".

Historically, two alternative explanations were offered for the occurrence of a coronal shock. Waves launched by an explosion may subsequently travel away from the site of their generation without further energy input. This type of shock is commonly called a blast wave, and the explosion at its origin is often thought to be the explosive energy release during a flare. Alternatively,

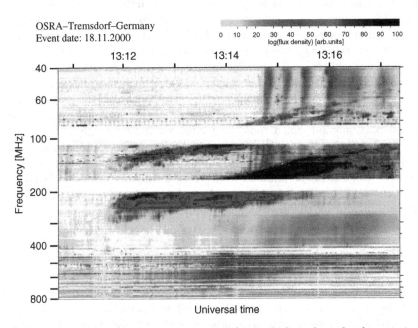

Fig. 2. Dynamic spectrogram of a radio event (*dark shading shows bright emission*) comprising a slowly drifting type II burst and fast-drifting type III bursts (Tremsdorf Observatory, Potsdam Astrophysical Institute)

shocks that are formed at the interface of a propagating plasma structure like a loop or a coronal mass ejection and the ambient plasma are referred to as "piston-driven" shocks. Discussions of these concepts are given in the chapters by Warmuth and by Mann & Vršnak.

Alexander Warmuth gives an extensive overview of large-scale perturbations, describing observations in the Hα and He I lines, at EUV and soft X-ray wavelengths, and in the radio domain. MHD waves, simple waves and shocks are briefly discussed. Warmuth concludes that the wave-like manifestations observed in different spectral ranges are physically related, and that the fastest perturbations reveal a common underlying disturbance with properties suggestive of MHD shocks. Alternative interpretations are also invoked, such as the idea that the "waves" could actually be signatures of plasma and magnetic field compression at the border of large-scale structures which are opened in the course of a CME. Such a scenario could account for EUV perturbations with irregular shapes or irregular propagation.

Shock waves are widely advocated as the accelerators of energetic particles in astrophysical plasmas. The Heliosphere is a particular laboratory where shocks and the associated waves and particle populations can be studied in detail, combining remote sensing and in situ measurements. In the solar corona shock waves may accelerate particles in flaring active regions (i.e. at the outflow jets of a reconnecting current sheet) or at remote sites. For example, fast coronal mass ejections are expected to drive shock waves through the corona and interplanetary space, and these shocks may accelerate energetic particles that are detected in space. The type III bursts emanating from the high-frequency type II band in Fig. 2 suggest that electron beams were accelerated at this coronal shock, as they are at the Earth's bow shock.

Nat Gopalswamy discusses the origin of large-scale coronal and interplanetary shock waves, as revealed by metric-to-kilometric type II burst emission, and their role in accelerating energetic particles that escape to interplanetary space. From statistical studies he concludes that the related shocks are all driven by fast coronal mass ejections. He further argues that large solar energetic particle events – i.e. conspicuous enhancements of energetic particle fluxes in space – are accelerated by such shocks. This is indicated by their association with fast and broad CMEs, and with type II bursts at decametre-to-hectometre wavelengths that show the presence of shocks while the CME travels through the high corona and interplanetary space. In Gopalswamy's view the combination of a high Alfvén speed in an active region and the acceleration of a CME in the corona could explain the formation of a shock at relatively high coronal altitudes, leading to a preferred correlation of fast CMEs and type II emission at relatively long wavelengths, and to the absence of metric type II emission (i.e. emission from coronal heights below $1\,R_\odot$) in some fast CMEs. Further statistical investigations suggest that particle fluxes in space are enhanced when a fast CME interacts with a previous slower CME. Gopalswamy discusses possible physical mechanisms that could explain this. An important conclusion from these investigations is the role of shock structure and of seed populations for the efficiency of shock acceleration.

The best studied example of a large-scale shock wave in the Heliosphere is the Earth's bow shock. It is collisionless and therefore produces a variety of signatures of instabilities, plasma waves and non thermal particle populations. David Burgess gives an overview of recent observations of particles, especially electrons, waves and radio emission, emphasising results from the four-spacecraft Cluster mission and comparing them with numerical simulations. Diffusive particle acceleration in the quasi-parallel part of the shock (Fermi acceleration) is discussed. This is also often invoked as the accelerating process at travelling shocks related to CMEs. The observations show many features predicted by theory, such as the exponential falloff of the accelerated particle density with distance from the shock. As the shock normal approaches the direction perpendicular to the upstream magnetic field (quasi-perpendicular shock), particles are accelerated to ever higher energies by mirroring, but this concerns mostly the particles in the tail of the distribution function. This shows that the energy distribution of the seed population is important for the efficiency of particle acceleration. The importance of seed populations for the shock acceleration is outlined e.g. by hot flow anomalies, a feature related to the interaction of the bow shock with a tangential discontinuity in the solar wind. They show that shock acceleration should be expected to depend critically on detailed features of the shock structure. Numerical simulations show the importance of structuring of the shock for electron acceleration, be it through the non-stationarity, the rippling caused by waves, or the curvature of the shock front.

A major problem for the interpretation of remote sensing observations as well as for the understanding of particle acceleration in the corona is our ignorance of the distribution functions of the "quiet" plasma. The importance of seed populations for the efficiency of particle acceleration at shocks has been outlined. It is customary in the analysis of stellar atmospheres to suppose that the particle distribution functions are not far from maxwellians. Milan Maksimovic addresses the signatures of non-maxwellian electron distributions in the solar wind, i.e. in some sense the seed population of any particle acceleration process. Spacecraft measurements in the interplanetary medium often show that the distribution functions have superhot components or suprathermal tails. These are indeed expected because for particles at high speeds the interplanetary medium and large parts of the solar atmosphere are non-collisional. Maksimovic gives an overview on the measurements in the solar wind and then discusses how suprathermal tails can affect our understanding of the acceleration of the solar wind and the heating of the solar corona. For the time being we can only speculate on the particle distribution functions in the corona, but we know that they are crucial and a better understanding of non-maxwellian features will be essential for further progress in understanding the corona.

Gottfried Mann and Bojan Vršnak describe aspects of current research on CMEs and shock waves. They introduce elementary scenarios of magnetic field evolutions that can lead to a CME. Interesting new results include a much closer relationship between CMEs and flares, such as the hint of correlations between CME parameters (speed) and flare intensity (see also the discussion by Gopalswamy) or reconnection rate in the CME-associated flare. A crucial question here, how the speed relevant to the reconnection rate can be extracted from the observations, still proves somewhat controversial. So is the nature of coronal shocks. The chapter concludes, unlike Gopalswamy (see above), that not all coronal shocks seem to be CME-driven. Controversy also surrounds particle acceleration at coronal shocks, since case studies may show coronal shocks without conspicuous acceleration of electrons or protons (see also the discussion by Gopalswamy). This does not exclude that shock waves may be efficient accelerators in active regions, notably at the interface of reconnection jets and the ambient plasma. Radio evidence for acceleration at these "termination shocks" is abundantly discussed in this chapter, as well as other processes related to turbulence and magnetic field retraction (betatron acceleration) in freshly reconnected magnetic loops.

3 Waves in Coronal Magnetic Field Structures and Plasma Diagnostics

Interest in wave phenomena in coronal magnetic structures has renewed in recent years, particularly in response to EUV and X-ray observations. In radio astronomy the search for periodicities in the emission of active regions and flares has a long tradition which has been re-invigorated by new imaging observations. A plausible interpretation of oscillations on minute scales are MHD waves propagating in limited spatial structures, i.e. magnetic flux tubes. Waves are interesting for many reasons: they transport energy, possibly playing a role in heating the corona; in the conditions of a flare they may accelerate particles. They are a potentially powerful diagnostic of coronal magnetic fields. These may be inferred from radio observations, or from the Hanle effect in spectral lines, but such deductions depend most often critically on assumptions on the geometry used to correct for line-of-sight integration. Dispersion relations for waves in coronal structures depend on characteristic plasma parameters such as the Alfvén and sound speeds. Thus the observation and analysis of pulsations opens up a new and fruitful window on these parameters: *coronal seismology*.

Valeri Nakariakov and Alexander Stepanov present formalisms describing MHD wave propagation in magnetic flux tubes. The dispersion relations derived for different wave modes in such configurations are applied to solar observations to infer key plasma parameters such as the density, temperature and magnetic field strength. An alternative approach using electric circuit models is then described, where coronal magnetic structures are represented by analogues of circuits from classical electrodynamics. Both approaches describe oscillations in loop systems but their physical content is rather different.

MHD dispersion relations represent in the first instance the linear response of the medium and reduce in the large wavenumber limit to the familiar Alfvén etc. waves. Circuit models on the other hand represent a sort of oscillation that is intrinsically nonlinear. From the comparison of predicted and measured pulsation periods, the authors derive electric currents and energies involved in some flare events on the Sun and in the flare star AD Leo. In this view the oscillations are intrinsic to the energy release process, rather than a secondary response to energy released by other means.

Alexander Nindos and Henry Aurass give a detailed overview of ongoing research in the field of radio pulsations. Pulsations are reviewed in all ranges of the radio waveband. Quasi-periodic phenomena in metric radio emission have been known for some time, but can now be studied with more sensitive spectrographs and with imaging observations. Modulations of broadband continua such as fiber bursts or zebra patterns are supposed to be closely related with magnetic fields, and allow one to estimate values of the otherwise elusive magnetic field strength. But quasi-periodic features are shown to exist not only in flare-related emission, but also in radio emission of active regions, the so-called "slowly-varying" component of solar radio emission.

Oscillations in the radio range may result from a variety of physical processes. The authors discuss different kinds of interpretation, including MHD waves, but also kinetic plasma instabilities related to the acceleration and trapping of energetic electrons. The possibility of limit cycle in wave-particle interactions has been demonstrated theoretically, and oscillatory behaviour may be intrinsic to coherent radiation mechanisms.

4 Outlook

The contents of this book reflect key topics of contemporary research in the physics of the solar atmosphere. The relationship between flares and CMEs, i.e. the question how the Sun decides to partition energy between heating and particle acceleration on one hand, large-scale motions and ejections of magnetic structures on the other, is a field of vigorous research, where new observations are required. In the coming years a fleet of instruments on ground and in space will continue observations, but new diagnostics will be added: vector magnetic field measurements from space with Hinode, stereoscopic observations with STEREO that will allow us for the first time e.g. to observe CMEs and the underlying regions of the corona. The combination of radio imaging and spectrography remains a necessary complement because of the sensitivity to non-thermal particles and to the unique constraints it provides for coronal plasma parameters, including the magnetic field.

Narrow-band and short-lasting spectral features in the radio band give us information on processes far beyond the scales which can be directly imaged. In many respects, including the systematic measurement of magnetic fields in the corona and the localisation of sites of fragmented radio emission related to particle acceleration in flares, broadband spectral imaging from centimetre to metre wavelengths will be an invaluable tool by which radio techniques will further our understanding of the active solar atmosphere. This is the objective of the *Frequency Agile Solar Radio Telescope* (FASR, http://www.ovsa.njit.edu/fasr/). A radioheliograph at centimetre-to-decimetre waves is under construction in China. At long metre wavelengths, the multi-purpose *Low Frequency Array* (LOFAR) will provide a considerable extension of our ability to track electron beams and shock waves through the corona. Observations in the mm and sub-mm range have opened a new window on highly energetic flare electrons, and on the response of the deep atmosphere to heating. Exactly how to interpret these observations is not yet clear and we may expect major advances in understanding from both ongoing development of instrumental facilities and new theory driven by these new observations. Extremes of solar particle acceleration may be further probed in the forthcoming GLAST gamma-ray mission (http://glast.gsfc.nasa.gov/).

The high efficiency of flare electron and ion acceleration implied by X-ray observations remains a major unsolved problem. At least on some occasions ion acceleration seems to pose similar problems. Continuing interrogation of the accumulating RHESSI data may help to resolve this, together with the unique sensitivity to higher electron energies afforded by observations in the 10s of GHz range.

Introductory and complementary information on the subjects discussed here can be found in the books cited in the list of references below.

References

1. Aschwanden, M.: Physics of the Solar Corona - an Introduction. Springer Verlag, Berlin, Heidelberg, New York (2004)
2. Benz, A.O.: Plasma Astrophysics - Kinetic Processes in Solar and Stellar Coronae. Kluwer Academic Publishers, Dordrecht (1993)
3. Treumann, R.A., Baumjohann, W.: Advanced Space Plasma Physics. Imperial College Press, London (1997)
4. Klein, K.-L.: Energy Conversion and Particle Acceleration in the Solar Corona. Springer, Lecture Notes in Physics, Vol. 612, Berlin, Heidelberg, New York (2003)

Part I

Particle Acceleration During Flares

Magnetic Complexity, Fragmentation, Particle Acceleration and Radio Emission from the Sun

Loukas Vlahos

Department of Physics, University of Thessaloniki, 54124 Thessaloniki, Greece
vlahos@astro.auth.gr

Abstract. The most popular flare model used to explain the energy release, particle acceleration and radio emission is based on the following assumptions: (1) The formation of a current sheet above a magnetic loop, (2) The stochastic acceleration of particles in the current sheet at the helmet of the loop, (3) the transport and trapping of particles inside the flaring loop. We review the observational consequences of the above model and try to generalize by putting forward a new suggestion, namely assuming that a complex active region driven by the photospheric motions forms naturally a large number of stochastic current sheets that accelerate particles, which in turn can be trapped or move along complex field line structures. The emphasis will be placed on the efficiency and the observational tests of the different models proposed for a flare.

1 Introduction

Radio emission from solar active regions during flares is closely related with two factors, (a) the rate of electron acceleration before, during and after the impulsive phase of the flare, and (b) the topology of the magnetic field in the active region hosting the flare. None of the main "actors" responsible for the characteristics of the radio emission is well known and this makes the direct modeling and interpretation of the radio observations extremely difficult.

High energy particles are also responsible for hard X-rays and γ-ray bursts emitted from the same active region, and which are less dependent on the details of the magnetic field topology. The combined analysis of the high energy emission (including Hard X-rays, γ-rays and radio) from the same event is an extremely valuable tool for our understanding of the physical processes behind the flares. Unfortunately, the events covered simultaneously in all wavelengths are only a few, and no conclusive results can be drawn for the source of energetic particles or the topology of the magnetic fields.

We can then conclude from the above that the direct reconstruction of the velocity distribution and the magnetic topology from the data is not possible, which has led many researchers to very simplistic models (Maxwellian with

L. Vlahos: *Magnetic Complexity, Fragmentation, Particle Acceleration and Radio Emission from the Sun*, Lect. Notes Phys. **725**, 15–31 (2007)
DOI 10.1007/978-3-540-71570-2_2 © Springer-Verlag Berlin Heidelberg 2007

power law tails, simple magnetic loops, single current sheets, etc). The alternative road is to estimate the high energy emission from the proposed flare model (direct modeling) and then compare the results of the model with the available data. This road seems more straightforward, but there is a fundamental obstacle, the scales of the physical processes involved in the formation of the unstable magnetic topology responsible for the flare are very different from those responsible for the dissipation of the magnetic energy and the acceleration of high energy particles.

MHD models can follow successfully the large scale evolution of the magnetic field, but the dissipation of magnetic energy (responsible for flares) is not necessarily an MHD processes (resistivity plays a crucial role). The acceleration of particles is a kinetic phenomenon appearing on all scales. Therefore, all attempts made so far to explain the flare and the coronal mass ejection (CME) inside the framework of the MHD theory have failed to explain the high energy phenomena.

In the past, high energy particles (energies above 25 keV) were assumed to carry a small percentage of the energy released in explosive phenomena. Recent estimates show that this is not true and more than 40% of the energy released in solar flares is going to high energy particles [1, 2, 3, 4, 5]. This leads us to the conclusion that kinetic phenomena play a crucial role in the understanding of solar flares.

Hybrid codes, following both the evolution of the MHD and kinetic aspects of a flare on all scales (from meters to thousands of kilometers) are not feasible today and the detailed modeling of solar flares remains an open problem for the future generation of computers. We then conclude that, since both the reverse and forward modeling are not possible today, we will go ahead with simple scenarios (cartoons) for the processes which we believe to be active during flares and follow their implications.

Two broad classes of flare models are widely used today: (a) The **break-up** type of models, representing mainly the flares which show close association with large scale events leading to CMEs [6, 7, 8], and (b) the **loop** models, which explain better the compact and relatively small flares which are not associated with CMEs. The splitting of the physical processes in categories is a useful tool for detailed studies, but it may lead us to wrong conclusions when the magnetic topologies, where the flares start, are extremely complex.

The mechanisms for the acceleration of particles and their transport are different in the two models. The expected radiation signatures are also very different and it is worth reviewing briefly the main features.

Our goal in this review is to present the current status of the energy release processes for solar flares and the associated acceleration mechanisms and to sketch the expected radio emission. Moreover we introduce a **third type** of flare model, which is based on the complexity of the magnetic topologies and the fragmentation of the magnetic energy release, which is associated with the turbulent photospheric motions and which results in complex magnetic fields and the formation of many current sheets of all scales.

In Sect. 2, the main "traditional" models (break up and loops) for the energy release and the expected acceleration of particles will be sketched. In Sect. 3, the energy release and the acceleration of particles in more realistic complex magnetic topologies will be analyzed and in Sect. 4 the expected radio emission from all three types of models will be discussed briefly. Our results will be summarized in Sect. 5.

2 Classical Models for Energy Release and Particle Acceleration

2.1 The Break-up Model for Flares/CME

A large variety of models demonstrates the connection between the flare and the CME [9, 10, 11, 12, 13, 14, 15]. All the above models start from simple magnetic topologies (2D or 3D) which are driven to instability by the shear along or perpendicular to a neutral line or by emerging magnetic flux. The schematic view of the models emerging from the proposed scenarios and corresponding simulations are shown in Fig. 1. There are several variants of the break-up model, according to the details of the initial magnetic field and the photospheric motions, these details are though beyond the scope of this article (see [16]).

The cartoon presented in Fig. 1 (top) shows that the high energy particles are covering a small portion (labeled with red/light gray) of the volume covered by the unstable structure. There are several acceleration regions in this model: (1) the current sheet, (2) the turbulent outflows, (3) the slow and fast shocks. In Fig. 1 **(bottom)**, the emphasis is given on the presence of several shocks surrounding the long and thin current sheet above the closed magnetic topology.

We could point out several weak points of the break up model, we will though focus our attention on the so called "number problem". It is well known that during a flare the required rate of particles accelerated is 10^{37} particles/sec, so for a flare lasting 100 secs more than 10^{39} particles will be accelerated. Translating this number to coronal conditions (mean density $10^9 - 10^8$ particles/cm^3), we conclude that the current sheet and the surrounding parts (fast jets, shocks etc) (see Fig. 1a) should cover a volume comparable to 10^{28}cm^3. Assuming that the thickness of the current sheet is several ~ 10 km we can reach the conclusion that the current sheet should be huge (10^{11}cm $\times$ 10^{11}cm) and should remain stable for 100s of secs. The plasma inside this volume should be replenished and accelerated to high energies with extraordinary efficiency. We believe that it is hard to prove that this huge structure will remain stable and active for so long inside an unstable magnetic topology. The break up of the current sheet and the formation of several fragments will be a natural consequence [17].

The 3D evolution of a simple bipolar photospheric magnetic field leads also to the break up model, but the magnetic topology is extremely complex [15].

CME/Flare Loop Structures

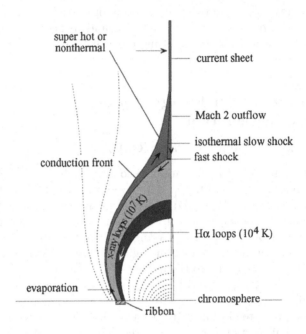

super hot or nonthermal

current sheet

Mach 2 outflow

isothermal slow shock

fast shock

conduction front

X-ray loops (10^7 K)

Hα loops (10^4 K)

evaporation

chromosphere

ribbon

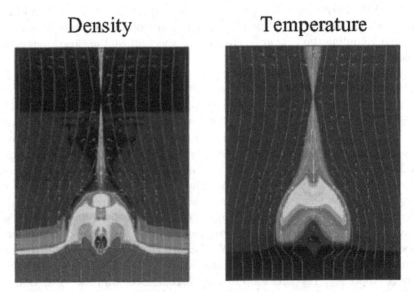

Density

Temperature

Fig. 1. The break up model. (**a**) The schematic representation [11], (**b**) A 2-D simulation [12]

The formation of a large number of stressed magnetic field lines (see Fig. 2) with numerous current filaments and fragments is apparent and can be the answer to the high energy emission observed. The simple magnetic topology for the current sheet, presented earlier and the associated simple accelerators (Direct E-field, constant flows and shocks) are replaced with much more complex topologies, which host a variety of accelerators in the 3D configurations.

The most prominent accelerators in the stressed 3D topologies mentioned above are the stochastic electric fields due to the sheared field lines. The expected electric fields are

$$E(r, t) = -V(r, t) \times B(r, t) + \eta J(r, t) \qquad (1)$$

where $V(r, t)$ is the plasma velocity, η is the resistivity, $B(r, t)$ the magnetic field and $J(r, t) \sim \nabla \times B(r, t)$ the current. In the 3D representation of the break up model the monolithic current sheet disappears and new, more advanced and interesting models for particle acceleration appear. We will come back to these models in the next section.

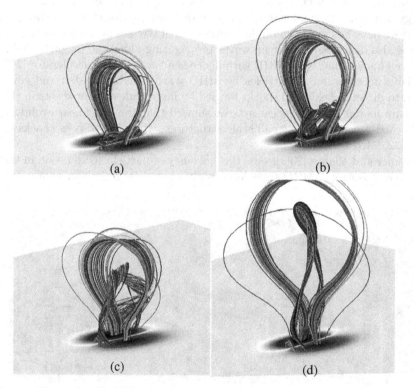

Fig. 2. Using the 3D MHD equations, even by starting from a simple magnetic geometry, the loop is led to a break up state with a very complex magnetic topology. The formation of numerous current sheets on all scales is apparent [14, 15]

2.2 The Loop Model

Nordlund and Galsgaard [18] solved numerically the non-ideal MHD equations to follow the evolution of photospheric stresses on a simple magnetic loop. The magnetic field initially was assumed to be uniform and anchored in the photosphere. The boundary velocity consists of a sinusoidal shear with a wave length equal to the length of the boundary. The orientation of the shear, the phase, the velocity amplitude and the duration of the individual driving events are random, with various limitations on the parameters [19].

As the boundaries of the loop are stressed by the randomly changing boundary flows, the loop is stressed into a state where numerous Unstable Current Sheets (UCS) are formed and distributed along the length of the loop (see Fig. 3).

Turkmani et al. [20] used the above model to study its efficiency to accelerate particles. A snapshot of the coronal magnetic field was used. The 3D structure of the electric field was estimated with the use of Eq. 1 (see Fig. 4). The scattered electric field accelerates ions and electrons in very short times ($\leq$ 0.1 sec). The electrons and ions are stochastically accelerated forming energy distributions similar to the ones needed to reproduce the observations. So far we have analyzed the random formation of stresses inside the loop. The loop is also disturbed by Alfvén waves propagating along the magnetic field. It is well known that weak MHD turbulence is a very efficient accelerator [21]. Diffusive acceleration of particles by MHD waves was contrasted and compared to direct E-fields and shocks formed by large scale current sheets in the break-up model. Several recent articles showed that the non linear evolution of the MHD waves forms small scale structures, which act also as shocks or UCS [22, 23, 24].

Arzner and Vlahos [23] discuss the efficiency of particle acceleration in the presence of isotropic MHD turbulence with anomalous resistivity as a proxy

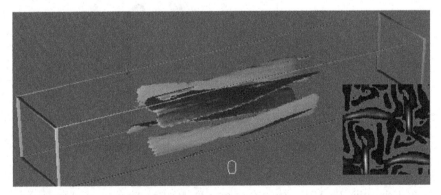

Fig. 3. The loop is stressed by random photospheric flows and is led to a state where numerous current sheets are present. A vertical cross section through the middle of the loop shows the formation of current sheets [18]

Fig. 4. The resistive electric field within the coronal loop, as calculated by the MHD model [20]

for the solar corona. The model for the MHD turbulence was relatively simple. They assume that the vector potential A was a superposition of Alfvén waves propagating along the external magnetic field B_0

$$A = \sum_k a(\mathbf{k}) cos(\mathbf{k} \cdot \mathbf{x} - \omega(\mathbf{k})t - \phi_k)$$

in axial gauge, $a(\mathbf{k}) \cdot \mathbf{v}_A = 0$ and with the dispersion relation $\omega(\mathbf{k}) = \mathbf{v}_A \cdot \mathbf{k}$, which is an exact solution of the induction equation with constant velocity $\mathbf{v}_A$. The $A(\boldsymbol{x}, t)$ is taken as Gaussian with random phases ϕ_k and (independent) Gaussian amplitudes $a(\mathbf{k})$, with zero mean and variance.

They analyze the evolution of a collisonless test particle in evolved homogeneous MHD turbulence with electromagnetic fields

$$B = \nabla \times A$$

$$E = -\partial_t A + \eta(J)J$$

where $\mu_0 J = \nabla \times B$ and the resistivity switches on to anomalously high values when the current exceeds a critical value

$$\eta(J) = \eta_0 \theta(|J| - J_c)$$

where $\theta(x)$ is the step function. The wave vector represents the random fluctuations along the external magnetic field. The formation of UCS inside the 3D topology of the magnetic field is a consequence of the non linear interaction of MHD waves with the plasma (see Fig. 5a).

Particles crossing the localized UCS will experience a sudden acceleration (or deceleration) (see Fig. 5b). These jumps are random and their characteristics are shown to be beyond the quasilinear analysis described by the standard

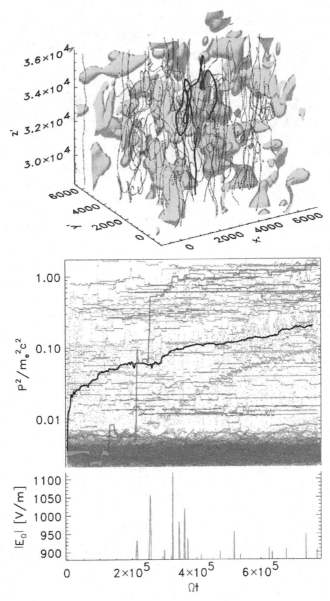

Fig. 5. (a) Location of the UCS. (b) Evolution of the electron momentum. Samples of trajectories show that the particles perform random walks and a few particles undergo very fast acceleration. All particles visit a sample of UCS [23]

Fokker-Planck equation [25]. We can then conclude that for high amplitude MHD waves the simple division between waves, shocks and large scale UCS is lost, and it is replaced with particle acceleration in a mixture of waves, UCS and shocks.

3 Energy Release and Particle Acceleration in Complex Magnetic Topologies

The initial magnetic topologies used so far in all the models discussed above were relatively simple bi-polar regions. The initially stable magnetic topologies were forced to instability by the continuous or random stressing from the photospheric motions. The evolution of the large scale instability led to the formation of a complex, fragmented structure.

The next level of complexity is to use a realistic magnetic field topology borne out from the linear and non linear force free extrapolation of the observed photospheric magnetic fields.

Vlahos and Georgoulis [26] use as the starting point of their analysis the magnetograms from a non flaring active region. Using the simplest possible method for the force free extrapolation [27], they determined the 3D magnetic field topology inside the active region. The real magnetic topologies are even more complex than the magnetic fields predicted by the linear force free state, but for the statistical analysis presented in their article the linear force free extrapolation is probably suitable. Using simple criteria for the potentially unstable currents, e.g the angular difference between two adjacent magnetic field vectors, B_1 and B_2 to exceed a certain value, since the steep magnetic field gradients favor magnetic reconnection in 3D magnetic topologies [28], they were able to define the location of the UCS. They concluded that active regions form naturally UCS even during their formation stage (Fig. 6). The free energy available in these unstable volumes follows a power law distribution with a well defined exponent (Fig. 7) [26]. We can then conclude that active regions store energy in many unstable spots forming UCS of all sizes. The UCS are fragmented and distributed inside the global 3D structure.

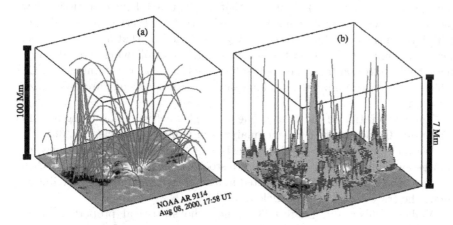

Fig. 6. (a) Linear force free field extrapolation in NOAA AR 9114, (b) Lower part of the active region atmosphere. Shown are the magnetic field lines (*red or light gray*) with the identified discontinuities for critical angle 10° [26]

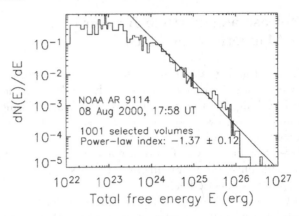

Fig. 7. Typical distribution function of the total free energy in the selected volume, on using a critical angle 14° [26]

The only approach which is capable to capture the full extent of the interplay of highly localized dissipation in a well-behaved large scale topology ("sporadic flaring") is based on a special class of models which use the concept of Self-Organized Criticality (SOC) [29]. The main idea is that active regions evolve by the continuous addition of new or the change of existing magnetic flux on an existing large scale magnetic topology, until at some point(s) inside the structure magnetic discontinuities are formed and the currents associated with them reach a threshold. This causes a fast rearrangement of the local magnetic topology and the release of the excess magnetic energy at the unstable point(s). This rearrangement may in turn cause the lack of stability in the neighborhood, and so forth, leading to the appearance of flares (avalanches) of all sizes that follow a well defined statistical law [30, 31, 32], which agrees remarkably well with the observed flare statistics [33]. We can then conclude, after many years of studies, that a possible model for the dynamic evolution of the active region is the following: The 3D magnetic field is stressed from the photospheric motions, forms continuously UCS which relax, re-arranging the local magnetic field and causing flares of all sizes (Fig. 8).

The acceleration of particles inside a complex active region being in Self-Organized Critical state has been analyzed in numerous articles [35, 36, 37].

We now pose a new question: Can the UCS become the local nodes for a large evolving network and accelerate stochastically electrons and ions? In this case the accelerator is not located in a single volume but it is distributed along the trajectory of the particle (see Fig. 9).

Vlahos, Isliker and Lepreti [37] study the statistical properties of an ensemble of isolated UCSs, and investigate the statistics of the energy gain when an entire distribution of particles moves through spatially distributed UCSs, all particles having random initial conditions. This question belongs to

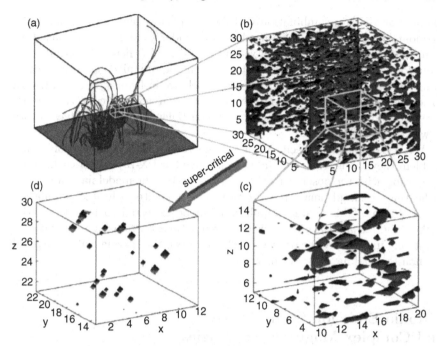

Fig. 8. (a) Simulated magnetogram of a photospheric active region and force free magnetic field lines, extrapolated into the corona (see [34]), (b) Subcritical current isosurfaces in space, as yield by the X-CA model [32], a particular SOC model, which models a subvolume of a coronal active region. (c) Same as (b), but zoomed. (d) Temporal snapshot of the X-CA model during a flare, showing the spatial distribution of the UCS inside the complex active region [37]

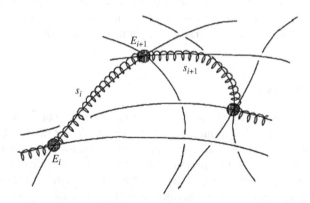

Fig. 9. A particle (*spiraling line*) follows the magnetic filed lines (*solid lines*), travels freely a distance s_i until it enters a UCS (*filled circle*) where it is accelerated by the associated effective DC field E_{i+1}. After the acceleration event the particle again moves freely till the particle meets the next acceleration event [37]

the field of MHD in combination with kinetic plasma physics (in what refers to anomalous resistivity). Also needed is an understanding of the spatial organization of an ensemble of co-existing UCSs and of their connectivity and evolution. A first hint to how the UCS might be organized spatially comes from the cited inquiries of SOC models, which are in favor of a global fractal structure with dimension around 1.8. The problem actually concerns the nature of 3-D, large scale, magnetized MHD turbulence, and it involves theory as well as observations.

With the concrete specifications of the random walk to the solar flare problem they made, they were able to achieve hard X-ray spectra which are compatible with the observations. Important is that the model naturally leads to heating of the plasma, or, more precisely, it creates a heated population in the plasma. This heated population can be expected to heat the entire background plasma through collisional interactions on collisional time-scales, explaining in this way the observed delay between the thermal soft X-ray and the non-thermal hard X-ray emission.

4 Radio Emission from Simple and Complex Magnetic Topologies

We are now ready to pose a very important question: Is it possible to identify the concrete radio signatures from the complexity of the magnetic field lines, the energy release and particle acceleration? The exact modeling of the radio emission from the structures presented above is still lacking but we can make several preliminary comments, hoping that both the new development in the theory and more importantly the new data expected from the Frequency Agile Solar Radiotelescope (FASR) [38] will give a new input to the analysis presented in this review. Let us discuss separately the well known parts of the flare related radio emission, starting from the microwave bursts.

Microwave Bursts

Microwave bursts are currently interpreted as the signature of mildly relativistic electrons trapped inside a magnetic loop [39, 40]. Acceleration of electrons and ions inside the loop can be a natural explanation for several well known high energy emissions: (1) The precipitating, mildly relativistic electrons produce the hard X-ray bursts forming the well known foot point emission, (2) the trapped mildly relativistic electrons produce the microwave bursts, forming large scale sources at the "loop top", (3) the precipitating relativistic electrons and ions are responsible for the γ-ray bursts.

Trapped and precipitating populations will also be present in more complex magnetic topologies. The fragmented energy release inside the loop will accelerate the electrons to almost relativistic energies [20]. The accelerated particles will cover large distances (thousands of kilometers) in relatively short time scales (fraction of a second), therefore microwave emission is not fragmented because the particles fill in a short time large parts of space, having moved away from the acceleration regions.

Radio instruments able to map the flaring region on a fast time scale (sub sec) can probably record, at the start of the flare, many small sources (dm spikes?), but eventually this will give away to a large and almost uniform source at later times.

The simple version of the break-up model will easily provide the precipitating particles and explain the Hard X-rays and γ-ray bursts but the trapping of particles inside the closed loop below the helmet is difficult to explain (cross field diffusion is rather difficult for these energies). Therefore we should expect a relatively long delay (several minutes) between microwave bursts and hard X-rays (such delays have not been recorded). The 3D analog of the break-up model leaves more room for precipitating and trapped particles with much less delay. Nobody has attempted so far to follow particles in a 3D magnetic topology resulting from the break-up model and hosting many reconnection sites.

We can then conclude that hard X rays, microwaves and γ-rays can easily be accommodated from the third type of flare models, which incorporates the fragmented energy release. We predict that the new generation of radio instruments will record many isolated microwave sources at the start of the flare.

Dm Spikes, Type III Bursts

The simple versions (assuming monolithic current sheets) of both flare models (loop model and the break-up) cannot account for the above bursts. Type IIIs appear usually in groups (isolated type IIIs are rare) at the rising phase of a flare.

These bursts can be explained from the fragmentation of the energy release in realistic magnetic topologies reconstructed from the extrapolation of the observed photospheric magnetic fields. Both types of magnetic field lines are present (closed and open, see Fig. 6). Therefore complex magnetic topologies and fragmentation are probably the explanation for the groups of type IIIs and dm spikes (see more in [41]).

A stochastic model for type III bursts was introduced and compared with observations [42]. In this model the active region is assumed inhomogeneous with a very large number of fragmented energy release regions (UCSs) connected to magnetic fibers. At the base of the magnetic fibers, random energy release events take place, in the course of which electrons are accelerated, travel along the fibers and eventually undergo bump-on-tail instability. Their main conclusion was that the observations are comparable with this model (see Fig. 10).

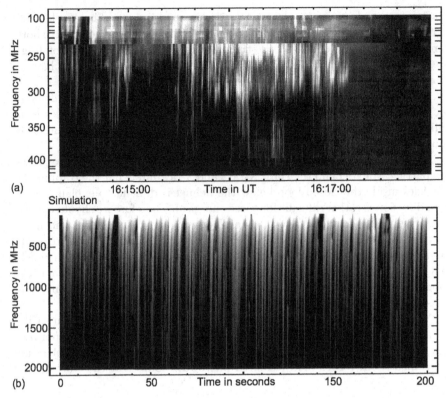

Fig. 10. (a) Spectrogram of type III event on 1980/06/27, 16:14:18 UT (time resolution 0.1 sec, shown duration 200 secs (b) 200 secs of a spectogram generated by the model with a time resolution 0.02 sec and frequency resolution 55.9 MHz. [42]

Type II Bursts

The break up model can easily account for the type II emission. It predicts several shocks, traveling in all directions, but type II bursts (forward and reverse) are not so common during flares. We can then ask: Why are type II bursts not always part of the flare/CME complex? The loop model on the other hand is not an efficient source of type II bursts.

Emission Before and After the Flare

The 3D version of the break up model can explain the pre-event radio emission as an expression of the build up of stresses leading to the instability, but it has difficulties to explain the long lasting (sometimes lasting for days) emission after the flare. The compact flare model can explain both since the loop is constantly under stress and the flare is a stronger explosion out of a series of explosions of all sizes. The fragmented complex loops model can explain these emissions much more easily. The realistic magnetic topologies easily predict both types of activity (pre flare and post flare). It is a challenge for the new radio instruments to investigate deeper this part of the flare problem.

Type I Bursts and Noise Storm Continuum

Type I bursts and noise storm continua can be the result of fragmentation of the energy release in large scale coronal magnetic complexes. The fragmented acceleration sources are responsible for the type I bursts. We propose that type I bursts are closely connected with the UCSs in the upper corona. Particle acceleration from the ensemble of the UCSs and subsequent trapping are responsible for the noise storm continuum. The type I bursts/noise storm continuum are analogous to the dm spikes/microwave bursts for the upper corona. The fundamental difference between the two types of bursts is that the dm spikes/microwaves are powered by a flare and the type I bursts/noise storm continuum are related with the micro flares appearing in the upper corona.

5 Summary

We have suggested in this review that flare models can be split into three very broad classes. The **break-up** model is responsible for the flares associated with CMEs and the **loop** model is connected with the compact flares. The complex model is an extension and generalization of the two classical models.

In 3D simulations of the break up model and of a randomly stressed loop the initial simple magnetic topology is forced to create many reconnection sites, and large scale current sheets collapse into many fragments.

Acceleration of particles is much simpler in an environment of fragmented energy release since the presence of stochastic E-fields, appearing in stressed magnetic topologies, naturally produces many UCS which collectively act to accelerate particles.

The next step in the development of realistic models is to consider the loop model and the break-up model in magnetic topologies borne out from the observed photospheric magnetic topologies (using linear or non linear force free extrapolation as the basic tool). In these cases the sharp division between the break-up and the loop flare model starts to disappear and a third type of model based on the fragmentation of energy release emerges, as shown here.

Current observations give only partial support for the break-up or the loop flare model. There are observations fitting naturally in the one or the other model and others which are hard to fit in any. We believe that the extrapolated magnetic fields, stressed by turbulent photospheric motions will create a mixture of closed and open field lines populated by randomly placed E-fields and they will be able to model the known observations. These topologies have the following characteristics:

- they are efficient accelerators;
- they have a mixture of open and closed magnetic filed lines;
- they can explain most types of bursts.

We believe that soon the next generation of flare models will emerge, where the current sheets will be hosted in a mixture of open and closed field lines. Forward modeling of the well known bursts inside these topologies will be an important diagnostic tool that will allow comparison to results from the radio instruments which are currently under development.

Acknowledgements

This work was supported in part by the Research Training Network (RTN) 'Theory, Observation and Simulation of Turbulence in Space Plasmas', funded by the European Commission (contract No. HPRN-eT-2001-00310). I would like to thank my colleagues, Drs Heinz Isliker, Manolis Georgoulis, Anastasios Anastasiadis, Kaspar Arzner, Rim Turkmani, Klaus Galsgaard, Bernhard Kliem and Prof. Peter Cargill for discussing and developing together several ideas presented in this review.

References

1. R.P. Lin, H.S. Hudson: Solar Phys **50**, 153 (1976)
2. B.R. Dennis, R.A. Schwartz: Solar Phys **121**, 75 (1989)
3. A.O. Benz et al.: Solar Phys **153**, 33 (1994)
4. P. Saint-Hilaire, A.O. Benz: Solar Phys **210**, 287 (2002)
5. G.D. Holman et al.: ApJ **595**, L97 (2003)
6. E.R. Priest, T.G. Forbes: Astron. Astrophys. Rev. **10**, 313 (2002)
7. B.C. Low: JGR **106**, 25141 (2001)
8. J.A. Klimchuk: in "Space Weather", P. Song, H.J. Singer and G.L.Siscoe (eds), AGU Geophysical Monograph, 125, 143, (2001)
9. R.A. Kopp, G.W. Pneuman: Solar Phys **216**, 123 (1977)
10. J. Heyvaerts, E.R. Priest, D. Rust: ApJ **50**, 85 (1976)
11. T.G. Forbes, E.R. Priest: ApJ **446**, 377 (1995)
12. T. Yokoyama, K. Shibata: ApJ **549**, 1160 (2001)
13. S. Antiochos, C.R. DeVore, J.A. Klimchuk: ApJ **510**, 485 (1999)
14. I.I. Rousev et al.: ApJL **588**, L45 (2003)
15. T. Amari, J.F. Luciani, J.J. Aly, Z. Mikic and J. Linker: ApJ **595**, 1231 (2003)
16. M. Aschwanden: *Physics of the solar corona*, (Springer, Berlin Heidelberg New York 2004)
17. B. Kliem, M. Karlicky, A.O. Benz: AAp **360**, 715 (2000)
18. A. Nordlund, K. Galsgaard: Tech. Rept., Astr. Obs., Copenhagen Univ. (1997)

19. K. Galsgaard, A. Nordlund: JGR **101**, 13445 (1996)
20. R. Turkmani et al.: ApJ, **620**, L59 (2005)
21. J.A. Miller et al.: JGR **102**, 14631 (1997)
22. P. Dmitruk et al.: ApJ **617**, 667 (2004)
23. K. Arzner, L. Vlahos: ApJL **605**, L69 (2004)
24. S. Moriyasu et al.: ApJL **601**, L107 (2004)
25. K. Arzner, L. Vlahos, B. Knaepen, N. Denewet: Springer Lecture Notes in Computer Science 3723 (Springer, Berlin 2005), 538
26. L. Vlahos, M. Georgoulis: ApJL **603**, L61 (2004)
27. C.E. Alissandrakis: AAp **100**, 197 (1981)
28. E. Priest, G. Hornig, D.I. Pontin: JGR **108**, 1285 (2003)
29. P. Bak, C. Tang, K. Wiesenfeld: Phys. Rev. Lett. **59**, 381 (1987)
30. E.T. Lu, R.J. Hamilton: ApJ **380**, L89 (1991)
31. L. Vlahos, M. Geopgoulis, R. Kluving, P. Paschos: AAp **299**, 897 (1995)
32. H. Isliker, A. Anastasiadis, L. Vlahos: AAp **377**, 1068 (2001)
33. N.B. Crosby, M.J. Aschwanden, B.R. Dennis: Solar Phys **143**, 275 (1993)
34. T. Fragos, E. Rantsiou, L. Vlahos: AAp **420**, 719 (2004)
35. A. Anastasiadis, M. Georgoulis, L. Vlahos: ApJ **489**, 367 (1997)
36. A. Anastasiadis, C. Gontikakis, N. Vilmer, L. Vlahos: AAp **603**, **422** 323 (2004)
37. L. Vlahos, H. Isliker, F. Lepreti: ApJ **608**, 540 (2004)
38. T. Bastian: Adv. Space Res. **32**, 2705 (2003)
39. M.R. Kundu, L. Vlahos: Spa. Sci. Rev. **32**, 405
40. T. Bastian: Proc-2000-Murdin, Vol. 3, 2553 (2000)
41. A. Benz: Solar and Space Weather Radiophysics, (eds D.E. Gary and C.U. Keller) Kluwer Academic Press, 203-221 (2004)
42. H. Isliker, L. Vlahos, A.O. Benz, A. Raoult: AAp **336**, 371 (1998)

Review of Selected RHESSI Solar Results

Brian R. Dennis[1], Hugh S. Hudson[2], and Säm Krucker[2]

[1] NASA Goddard Space Flight Center, Solar Physics Branch, Code 612.1,
  Greenbelt, MD 20771, USA
  `Brian.R.Dennis@nasa.gov`
[2] Space Sciences Laboratory, University of California, Berkeley, CA 94720, USA
  `hhudson@ssl.berkeley.edu krucker@ssl.berkeley.edu`

Abstract. We review selected science results from RHESSI solar observations made since launch on 5 February 2002. A brief summary of the instrumentation is given followed by a sampling of the major science results obtained from the soft X-ray, hard X-ray, and gamma-ray energy domains. The thermal continuum measurements and detection of Fe-line features are discussed as they relate to parameters of the thermal flare plasma for several events, including microflares. Observations of X-ray looptop, and rising above-the-loop sources are discussed as they relate to standard models of eruptive events and the existence of a current sheet between the two. Hard X-ray spectra and images of footpoints and coronal sources are presented, showing how they can be used to separate thermal and nonthermal sources and determine the magnetic reconnection rate. Gamma-ray line images and spectra are presented as they relate to determining the location, spectra, and angular distribution of the accelerated ions and the temperature of the chromospheric target material. Finally, we discuss the overall energy budget for two of the larger events seen with RHESSI.

1 Introduction

With its broad energy coverage from 3 keV to 17 MeV, the Reuven Ramaty High Energy Solar Spectroscopic Imager (RHESSI) is providing definitive observations of the three major components of solar flares: plasma at ≥ 10 MK, electrons accelerated to ≥ 10 keV, and ions to ≥ 1 MeV. During a large flare, thermal bremsstrahlung from heated plasma dominates the observed soft X-ray emission to energies often as high as a few tens of keV; nonthermal bremsstrahlung continuum from accelerated electrons is observed at higher, hard X-ray (and sometimes gamma-ray) energies; and line emission from nuclear transitions caused by accelerated ions is observed in gamma-rays from ~ 400 keV to ~ 8 MeV. In all of these spectral domains, RHESSI has superior capabilities compared to previous instruments. In each case, new and interesting results have already been obtained and many more are promised as the knowledge of the instrument improves, the analysis software is extended to

B. R. Dennis et al.: *Review of Selected RHESSI Solar Results*, Lect. Notes Phys. **725**, 33–64 (2007)
DOI 10.1007/978-3-540-71570-2_3   © Springer-Verlag Berlin Heidelberg 2007

fully exploit RHESSI's imaging spectroscopy capabilities, and observations of the over 14,000 flares recorded to date are fully interpreted.

The impact of the RHESSI observations has been greatly amplified by the contemporaneous observations made with the vast array of other solar instruments currently in operation. These include observations over a broad spectrum of wavelengths from soft X-rays, EUV, UV, and optical, to radio. They provide thermal, magnetic, and morphological context information that is indispensable for the interpretation of the RHESSI observations. Combined with the in situ particle and field measurements in the near-Earth environment, they also provide information on coronal mass ejections (CMEs), the other great energetic solar phenomenon that is often, but not always, associated with flares.

In this review, we summarize some of the early results obtained with RHESSI from its first two years in orbit. The paper is organized by energy into the basic domains of soft X-rays from about 3 to 20 keV, hard X-rays from about 20 to 400 keV, and gamma-rays from about 400 keV to 17 MeV. These roughly correspond to the energy ranges dominated by emissions from thermal plasma, nonthermal electrons, and nonthermal ions, respectively, but there are clearly some overlaps in these interpretations as will become evident in the different sections of the paper.

2 Instrumentation

As described more fully by [36], RHESSI is a single instrument mounted on a spinning spacecraft with the spin axis pointed close (within $\sim$10 arc minutes) to the center of the solar disk. The instrument consists of an X-ray/gamma-ray spectrometer that views the Sun through a set of modulation collimators. The spectrometer has nine cylindrical germanium detectors, each 7.1 cm in diameter and 8.5 cm long. They detect photons from 3 keV to 17 MeV with fine energy resolution varying from $\sim$1 keV (FWHM) at the low energy end to $\sim$4 keV at 2 MeV. Above each detector sits a modulation collimator made up of two identical grids separated from one another by 1.55 m. All the grids consist of parallel slats separated by slits of comparable width. All the grid slats are tungsten except for those on the finest grid pair, which are molybdenum. As the spacecraft rotates at $\sim$15 rpm, the modulation collimators convert the spatial information about the source that is contained in the photon arrival directions into temporal modulation of the germanium detector counting rates. Smith et al. [70] and Hurford et al. [26] describe how the resulting telemetered energy and timing information about each photon recorded in all nine detectors is used on the ground with specialized analysis software [63] to give RHESSI its imaging spectroscopy capabilities. Angular resolutions as fine as 2 arc seconds are possible, and sources as large as 180 arc seconds can be imaged anywhere on the solar disc and up to several arc minutes above the limb. Several innovations have been incorporated into the instrument design to increase the dynamic range of flare intensities that can be recorded with RHESSI so that weak microflares are detected at times of low activity while detector saturation and spectral distortion are minimized during the most intense gamma-ray line flares. These innovations include detector segmentation, movable shutters, and high-rate electronics with pile-up suppression [70].

3 Soft X-rays

The soft X-ray spectral range from RHESSI's low energy limit of ∼3 keV to about 20 keV is of great interest since, in many flares, it is the region of transition from thermal to nonthermal emission. Thermal bremsstrahlung continuum and line emissions are seen in this energy range from flare plasmas with temperatures as low as ∼7 MK and as high as many tens of MK. Studying this thermal component can reveal not just the total energy in the hottest flare plasma throughout the flare but also information about the composition of the plasma, its density, and possibly any departure from ionization equilibrium. Nonthermal X-ray emission with a flatter power-law spectrum ($\epsilon^{-\gamma}$ where ϵ is the photon energy) is also seen above ∼10 keV, particularly during the impulsive phase of a flare, from electrons accelerated to tens of keV and higher. Many observations have shown that these accelerated electrons carry a large fraction of the total energy released in many flares, thus heightening the interest in the nonthermal component. Because these accelerated electrons have a steep power-law energy distribution ($E^{-\delta}$ where E is the electron energy) with δ generally >2, most of the energy resides in the lower energy electrons. Indeed, for such a steep power-law spectrum, a lower energy flattening or cutoff must exist to keep the total nonthermal energy finite. The determination of the cutoff energy is critical for any evaluation of flare energetics.

The RHESSI instrument provides much higher sensitivity in the soft X-ray energy range above ∼3 keV than has previously been available. Earlier solar hard X-ray instruments, such as the Hard X-ray Burst Spectrometer (HXRBS) on the Solar Maximum Mission (SMM) and the Hard X-ray Telescope (HXT) on Yohkoh, had entrance windows that absorbed emission below ∼15–25 keV to avoid saturation from the intense thermal emissions in large flares. RHESSI accommodates medium and large flares by automatically inserting shutters (aluminum discs) in front of detectors to absorb low energy solar photons and hence avoid saturation. Thus, when no shutters are in the detector lines of sight to the Sun, RHESSI is about 100 times more sensitive than previous instruments at around 10 keV. Even with the shutters in front of the detectors during the largest flares, thin areas in the aluminum discs forming the shutters allow small fractions of the photons to pass through, making spectroscopy still possible down to about 5 keV.

3.1 Thermal Plasma

RHESSI observes emission from thermal plasma with temperatures above ~7 MK. This includes thermal free-free and free-bound continua plus line emission, primarily from highly ionized iron and nickel. The continuum, with its pseudo-exponential shape, allows RHESSI to provide accurate determinations of the emission measure $EM = \int N_e^2 \, dV$ (where N_e is the electron density and V is the emitting volume) and temperature T assuming an isothermal plasma. Differential emission measure analysis is also possible by combining RHESSI data with observations from other instruments in different wavelength ranges. With its ~1-keV FWHM energy resolution, RHESSI does not resolve the satellite-line structure of the Fe and Ni lines between ~6.4 and 10 keV described by [49]. Instead, it sees two broad Gaussian-like features when the plasma temperature is ≥10 MK, one centered at ~6.7 keV and the other weaker feature at ~8 keV. These features are made up of a large number of FeXXV lines and FeXXIV dielectronic satellites, with other lines due to FeXXVI and Ni XXVII contributing at higher temperatures. Such a spectrum is shown in Fig. 1, where data from the RESIK Bragg crystal spectrometer [75] on the Russian *CORONAS-F* spacecraft have been added to the spectrum

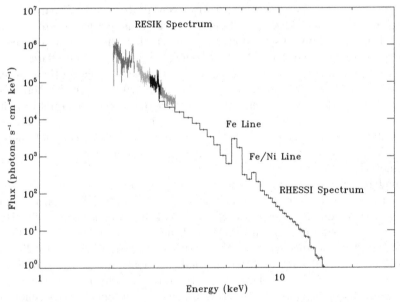

Fig. 1. X-ray spectra determined from RESIK and RHESSI observations for a time (03:00 UT) just preceding the main M2 flare on 2003 April 26. A temperature of 18.6 MK and an emission measure of 2×10^{47} cm^{-3} are obtained from a fit to the RHESSI spectrum assuming the existence of two line complexes with Gaussian profiles. (From [12])

derived from RHESSI observations for the early stages of an M2 flare seen by both instruments on 2003 April 26.

The two line features in the RHESSI spectra provide information on the plasma temperature and iron abundance that is independent of similar information derived from the continuum. A convenient way of expressing the intensities of the two line features is through the equivalent width, i.e. the energy width of a portion of the continuum at the line's energy with flux equal to that of the line feature. The variation of equivalent width with T is given in Fig. 2 (smooth curve) based on CHIANTI version 5 [13], with "coronal" element abundances ([Fe/H] $\sim 1.6 \times 10^{-4}$) and recent ion fraction calculations. This should be compared with values determined from RHESSI spectra similar to that shown in Fig. 1 for the flare on 2003 April 26. As can be seen, for observations in the A1 attenuator states (thin shutters in place over the detectors) on the rise and decay of the flare, the observed equivalent widths of the Fe-line feature lie close to the theoretical curve, giving support for a coronal abundance of Fe in the flare plasma. However, some of the equivalent widths measured in the A3 attenuator state (both thick and thin

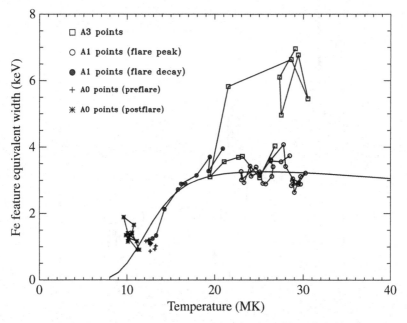

Fig. 2. Equivalent width (keV) of the Fe line feature at 6.7 keV plotted against T (*smooth curve*) as calculated from CHIANTI assuming coronal abundances, i.e. Fe/H and Ni/H equal to 4× photospheric, and [41] ion fractions. Observed *RHESSI* values in attenuator states A0, A1, and A3 for the M2 flare of 2003 April 26 are shown as points connected by lines (see legend for line styles and plotting symbols). (From [12])

attenuators in place) lie well above the predicted curve. This apparent discrepancy has been encountered in the spectral analysis of several other flares. It is possible that the presence of a multi-thermal or two-component emission measure spectrum contributes to this discrepancy or that plasma with a higher iron abundance appears at this time, but instrumental explanations for these anomalous variations are also under investigation.

Since both line features seen in RHESSI spectra are believed to be produced primarily from iron, it should be possible to obtain a measure of the plasma temperature independent of the iron abundance from the flux ratio of the two line features. Caspi et al. [7] analyzed RHESSI spectral observations for several flares and determined this line ratio as a function of the temperature derived from the continuum. The results for one flare are commpared in Fig. 3 with the Chianti-predicted variation in this line ratio with plasma temperature. There is general agreement between the measured and predicted values suggesting that this is a viable alternative method for determining the

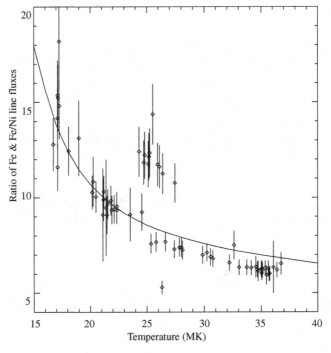

Fig. 3. Results of RHESSI spectral analysis showing the ratio of the Fe to Fe/Ni line fluxes at 6.7 and 8 keV, respectively, plotted as a function of the plasma temperature derived from the continuum assuming a single temperature (from [7]). The uncertainties on the temperatures are of the order of 1 MK. The *solid curve* is the predicted variation from CHIANTI assuming ionization fractions given by [41]. Data points are for the X4.8 gamma-ray line flare observed on 2002 July 23 starting at 00:18 UT in various attenuator states

plasma temperature. However, there are significant deviations from the predicted values at certain times during the flare. Analyses of other flares also show similar significant deviations during a flare and from flare to flare. These discrepancies may be the result of the multi-temperature nature of the flare plasma, but possible instrumental interpretations are also under investigation to explain these unexpected results.

3.2 Microflares

Figure 4 shows an example of RHESSI microflare observations (see also [15, 32, 39, 49]). Spectral investigations show the existence of a thermal and a non-thermal component (Fig. 5). The power-law fits ($\epsilon^{-\gamma}$) to the non-thermal component of the photon spectrum extend to below 7 keV with values of γ between 5 and 8. They imply a total non-thermal electron energy content of between 10^{26} and 10^{27} ergs [15, 32]. Except for the fact that the power-law indices are steeper than those generally found in regular flares, the investigated microflares show characteristics similar to large flares. Since the total energy in non-thermal electrons is very sensitive to the value of the power-law index and the energy cutoff, these observations will give us better estimates of the total energy input into the corona. In earlier work with observations above $\sim$25 keV, the cutoff energy was often set to 25 keV (e.g., [10]). For regular

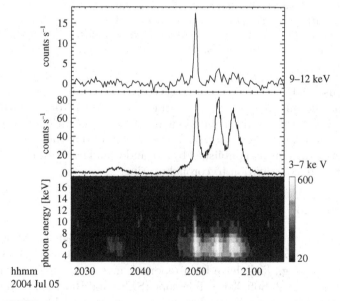

Fig. 4. RHESSI observations of a series of at least 5 microflares in the space of about 30 minutes on July 5, 2004. The *top* and *middle* time profiles are for the indicated energy ranges of 9–12 and 3–7 keV, respectively. The *bottom* plot is a spectrogram representation of the same data

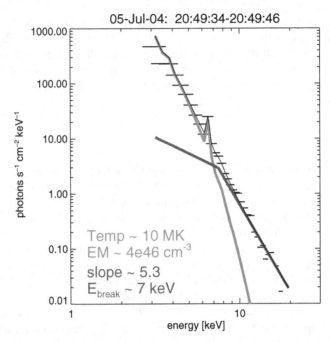

Fig. 5. Fitted spectrum of the hardest microflare shown in Fig. 4 at around 20:49 UT. A thermal (*red or dark gray*) and non-thermal (power-law) fit (*blue or black*) is shown. A spectral index γ of 1.7 was assumed below the $\sim$7 keV break energy. A fit with two thermal components gives a slightly worse, but still acceptable, fit. It gives a high temperature of $\sim$35 MK for the hotter component with an emission measure about 130 times smaller than that of the cooler component

flares, the use of 10 keV instead of 25 keV gives energies larger by a factor of $\sim$10. For the microflares present in this work, the factor is $\sim$500, since the spectra are steeper and the cutoff energy lower. Hence, the correction for smaller events seems to be larger. This would require renormalization of the flare frequency distribution published by [10] and would lead to a re-evaluation of the microflare contribution to coronal heating.

3.3 Evidence for a Current Sheet

Notwithstanding the observational problems for flare models based on large-scale magnetic reconnection (e.g., [25]), several pieces of indirect evidence for such models (e.g., [51]) have been reported, mostly using X-ray observations from the Yohkoh Soft X-ray Telescope (SXT) and Hard X-ray Telescope (HXT). Cusp-shaped soft X-ray flare loops were reported by [76] and [77], with high-temperature plasma along the field lines mapping to the tip of the cusp. However, [45] questioned the reality of the cusps in at least one flare. Tsuneta et al. [76] reported the expected increase of loop height and footpoint separation with time. Masuda et al. [40] discovered a hard X-ray source above the soft X-ray loops. Evidence has been presented for horizontal inflow above the cusp region by [80], downflow above the loop arcades by [42], and an upward-ejected plasmoid above the loops by [68] and [53].

New evidence for magnetic reconnection has been obtained with RHESSI observations. Sui & Holman [72] analyzed a series of flares in April 2002, all from the same active region, showing bright flare loops and coronal X-ray sources above them. Such an X-ray source structure is shown in Fig. 6, where RHESSI contours are overlaid on a contemporaneous TRACE image taken during the impulsive phase of one of these flares on 15 April. Both a bright loop and a source above the loop are evident in this image. The projected altitudes of these X-ray sources vs. time and energy are plotted in Fig. 7, along with the X-ray light curves in different energy bands. Initially, the centroid of the bright loop-top source appeared to move to lower altitudes at about ∼9 km s^{-1}. Similar initial decreases in apparent height have been reported for other flares [73]. The reason for this initial fall is not known or predicted by any of the reconnection models. Sui et al. speculate that it could be the result of the newly reconnected field lines relaxing to a near semicircular state. Alternatively, it could be support for implosion [23].

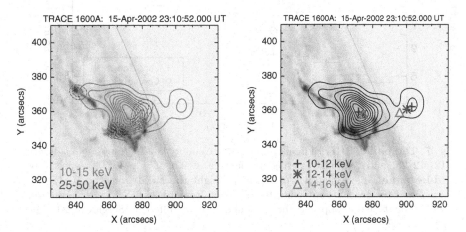

Fig. 6. Left image: TRACE 1600-Å image of the flare that started at 23:07 UT on April 15, 2002, overlaid with RHESSI contours for the indicated energy ranges. Note the loop-top structure in the 10–15 keV source (*red or dark gray contours*) with a separated source around [905/360] arcsec. HXR emission (*dashed blue or gray*) is seen from two footpoints ([840/370] and [865/350]), plus a third source around [880/355] that is most probably located in the corona. **Right image**: same as the left image but showing only the 10–15 keV contours. The centroids of the looptop and coronal sources are shown for the different energies as indicated. Note that the higher energies of the looptop source are at higher altitudes (1 Mm = 1.3 arcsec) whereas the higher energies of the coronal source are at lower altitudes. This suggests that the flare energy must have been released between the two sources. (After [72])

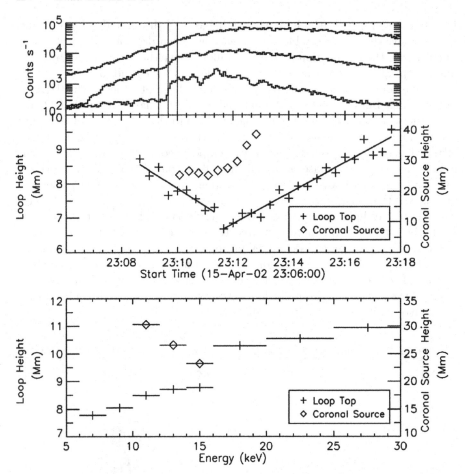

Fig. 7. Top panel: RHESSI light curves in three energy bands (*from top to bottom*): 3–12, 12–25, and 25–50 keV scaled by factors of 2.0, 0.5, and 1.0, respectively. **Middle panel**: Time histories of the loop height (obtained from the 10–12 keV images) and the coronal source height (obtained from 10–25 keV images). **Bottom panel**: Height of the loop and the coronal source at different energies at 23:11:00 UT. After [72]

After the impulsive rise, the upper part of the coronal source separated from the underlying flare loop. Sui and Holman [72] speculate that this is the result of the initial X-type magnetic configuration evolving into a current sheet with a Y-point at each end [51]. The X-ray bright underlying loops appeared to rise at ~8 km s^{-1} after the HXR emission had peaked. The separated coronal source was stationary at first but then moved out of the RHESSI field of view at ~300 km s^{-1}.

When this result was first presented, the reality of the relatively weak coronal source above the loop top was questioned since it was only ~20%

as intense as the looptop source, i.e., close to the current RHESSI dynamic range capability. However, TRACE 195Å difference images recently prepared by Veronig (private communication) for a similar flare on 16 April 2002 show a rapidly moving structure, co-spatial and co-temporal with the RHESSI coronal source, thus dispelling lingering doubts that this relatively weak source might be an artifact of the RHESSI Fourier imaging technique.

The remarkable feature of the loop-top source and the overlying coronal source reported by [72] is their oppositely directed temperature gradients as determined from the RHESSI images - the temperature of the underlying loops increased with apparent altitude whereas the temperature of the separated coronal source decreased with altitude. This is illustrated in Fig. 6 (right), where centroid locations are shown for the two sources at different energies during the peak in HXRs. This composite image shows that the highest temperatures are located in the regions where the two sources are closest to one another (Fig. 6 right). This effect can also be seen in the bottom panel of Fig. 7, where the apparent altitudes of the two sources are plotted vs. photon energy. Note the opposite dependence on energy of the looptop and coronal sources. Sui & Holman [72] interpret this as strong evidence that the energy must have been released between the two sources by magnetic reconnection in a current sheet. This results in the formation of new bright structures both above and below the current sheet. The new structures are hotter than those formed earlier since the latter cool rapidly by both conduction and radiation.

How the newly formed magnetic structures become filled with hot X-ray emitting plasma is not certain. In the classic flare model, the magnetic reconnection accelerates electrons that propagate down the field lines to the footpoints. There, they emit the observed HXRs and heat the chromospheric gas, which moves explosively back up (the so-called chromospheric evaporation), filling the loop with hot plasma. Feldman [20] has pointed out significant problems with this scenario and suggests instead that in situ heating of the gas in the current sheet produces the bright loop-top thermal sources. This would also explain the appearance of the hot source above the looptops, which is difficult to account for in the classic chromospheric evaporation model. Clearly, observations of other flares are needed to resolve these issues.

4 Hard X-rays

At energies above the thermal emissions, the hard X-rays provide the most direct information about the electrons accelerated during the flare. An early flare observed with RHESSI on 20 February 2002 provides a relatively simple example of the information that is available and how it can be interpreted [32, 2, 74]. RHESSI images at the HXR peak of this flare are shown in Fig. 8. They are interpreted as showing both thermal emission from a hot loop at low energies and nonthermal emission from footpoints at high energies. Even in this simple case, however, the situation is more complicated since a weak but significant source is seen closer to the limb, particularly in the 16–18 keV

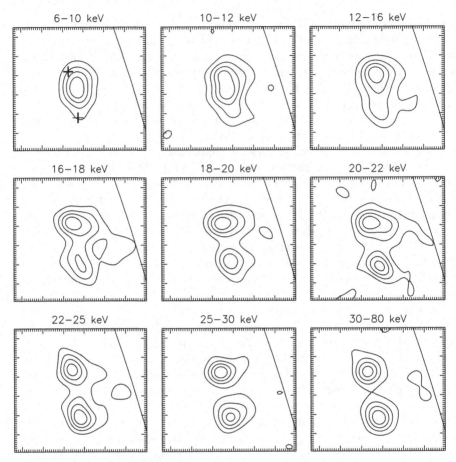

Fig. 8. RHESSI Cleaned images in different energy ranges between 6 and 80 keV for the flare on 20 February 2002 from 11:06:000 to 11:06:39.6 UT. The thermal source can be seen at low energies, below ∼15 keV, between the two footpoint sources that are clearly separated at higher energies. The contour levels are at 20, 40, 60, and 80% of the peak value in each image. The plus signs (+) in the 6–10 keV image mark the peaks of the two sources seen at 30–80 keV

image in Fig. 8. Sui et al. [74] interpreted this as a high altitude source around the top of a larger loop or arcade of loops.

The spatially integrated spectrum of this same flare on 2002 February 20 shown in Fig. 9 provides for a clean separation between the thermal and non-thermal emissions. It supports the view suggested by the images of Fig. 8 that thermal emission dominates below ∼10 keV. The nonthermal spectrum seems to extend down to ∼10 keV. This means that the energy in the accelerated electrons is significantly higher for this event than would have been estimated previously by arbitrarily assuming a lower cutoff energy of 20 or 30 keV. This

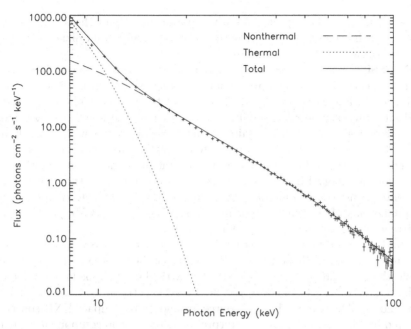

Fig. 9. RHESSI spatially integrated photon spectrum of the flare on 20 February 2002 for a 14-s time interval starting at 11:06:10 UT (after [74]). The indicated thermal and nonthermal curves give the best fit to the count-rate spectrum. The thermal continuum spectrum is for a temperature of 15 MK and an emission measure of 2×10^{48} cm^{-3}. The nonthermal spectrum is the photon spectrum that would be produced by a double power-law electron spectrum assuming thick-target interactions. The spectral index (δ) of the electron spectrum used for the fit is 4.4 below a break energy of 100 keV and 5.5 at higher energies

becomes increasingly important for steep nonthermal spectra and can have major implications in estimating the importance of the accelerated electrons in the overall flare energy budget as discussed below in Sect. 6.

The fitted spectrum shown in Fig. 9 was determined using a forward-folding method starting with a double power-law electron distribution having a low-energy cutoff. The resulting bremsstrahlung photon spectrum was computed assuming thick-target interactions. A thermal spectrum was then added to it and the combination folded through the RHESSI instrument response matrix to give a predicted count-rate spectrum. The parameters of the electron distribution and the thermal spectrum were then modified iteratively to minimize the value of χ^2 relating the predicted and measured count-rate spectra.

Other spectral analysis techniques have been tried to avoid having to make any assumptions about the form of the electron distribution. Kontar et al. [30] have used an inverse regularization method to analyze RHESSI data for a flare on 26 February 2002. They conclude that a suspected dip in the spectrum at

~20 keV and a similar dip at ~55 keV reported by [50] for the 23 July 2002 flare cannot be confirmed but must, at present, remain a tantalizing mystery (see also Brown et al., this volume).

RHESSI's imaging spectroscopy capability allows us to determine independent spectra for the spatially separated sources. This can be seen from the images in Fig. 8. Sui et al. [74] showed that the footpoint photon spectra can be fit with a power-law having $\gamma = 3$, whereas the looptop region has $\gamma = 4$. Krucker and Lin [32] obtain similar values for γ but with a break at about 20 keV early in the flare and softening with time after the peak. Aschwanden [2] were able to determine the height of the footpoint sources as a function of energy from the small arcsecond differences in the location of the source centroids in different energy bands. They found agreement with the predictions of the thick-target model assuming the very simple Caltech Irreference Chromospheric Model (CICM, [19]).

The high resolution and statistical accuracy of the energy spectra being measured with RHESSI allow more subtle effects to be investigated beyond the simple thermal and power-law spectra. In particular, the albedo component of the flux significantly modifies the measured spectrum, as pointed out by [1, 3] and [5]. The extended 'halo' source around more compact HXR sources reported by [61] and [62] may be interpreted as the albedo component. Other interpretations are still possible, however, such as thin-target or thermal loops, or other unidentified diffuse structures.

A second effect associated with the thick-target interpretation of the HXR emission is the varying ionization along the electron paths [6]. It is expected that the plasma will be fully ionized in the coronal magnetic loops but that the ionization level will decrease with depth in the chromosphere. The long-range collisional energy losses are reduced as the ionization level falls, resulting in a higher effective HXR bremsstrahlung efficiency. Consequently, the HXR emission at high energies (≥ 100 keV) is a factor of 2.8 more intense for electrons that penetrate down to the neutral gas target than for lower energy electrons that stop in the fully ionized plasma in the corona. The predicted HXR spectrum from a power-law electron injection spectrum of the form $E^{-\delta}$ (where E is the electron energy) has a power-law index of $\gamma = \delta - 1$ at both low and high energies but $\gamma < \delta - 1$ in between.

As shown in Fig. 9, RHESSI photon spectra often show deviations from a simple power-law in the range from 20–100 keV. Such spectral breaks may be associated with an acceleration process that gives corresponding breaks in the electron spectrum. However, [29] pointed out that they could be the result of the effects of changes in the ionization level across the transition region, and that the electron spectrum could still be a simple power-law as expected from some acceleration models. Other effects, such as the albedo flux discussed above and various instrumental effects, could also result in breaks in the photon spectrum that are not an indication of breaks in the electron spectrum. Important instrumental effects that remain uncertain are pulse pile up at counting rates above ~10,000 counts per detector (complicated by the image-dependent modulation of the counting rates), uncertainties in the instrument response matrix, and uncertainties in the background that must be subtracted from the measured count-rate spectrum. All of these effects are being actively investigated so that the full potential of the RHESSI observations can be achieved.

4.1 Separation of Thermal and Nonthermal Emission

Conventionally, one separates the HXR spectrum into a thermal and a non-thermal component. This is a most important, but often a very difficult task in the analysis of RHESSI observations. It is critical in determining the relative energies in the thermal plasma and in the accelerated electrons; this will be revisited in Sect. 6 on flare energetics.

Sometimes the separation of thermal and nonthermal emission can be relatively easy, as, for example, when we see hard X-ray emission from two bright sources that can be identified as the footpoints of magnetic loops or arcades. This is the case in the images of the 20 February 2002 flare shown in Fig. 8. Strong evidence that such footpoint emission is nonthermal comes from the simultaneity to within 0.1 s of the peaks from the two footpoints for the flares presented by [58] and [59] based on Yohkoh HXT observations. Convincing indirect evidence comes from the close association of HXR footpoint emission with low-temperature emissions such as the white-light continuum (e.g., [43]).

One can usually be safe in assuming that a coronal HXR source is thermal but this may not always be true. For example, both [33] and [78] have argued that coronal HXR sources seen in two different flares were, in fact, nonthermal.

The HXR time history in the impulsive phase is also an indicator of the thermal or nonthermal nature of the sources – the more impulsive emission is thought more likely to be nonthermal. This cannot be demonstrated unambiguously, however, but it is often a useful clue that can support or oppose assumptions made on the basis of other information.

Spectrally, there is often a clear distinction between the steep thermal component at lower energies and the flatter, power-law function at high energies, as shown in Fig. 9. Frequently, however, especially during the early stages of a flare, there is no clear spectral distinction, and a power-law with a single index can fit the data from the lowest energies covered by RHESSI up to the highest energies at which the flare emission is above the instrumental background level. Holman et al. [22] faced this problem in their analysis of the HXR spectra for the 23 July 2002 gamma-ray line flare. Before the main impulsive rise in HXRs, the count-rate spectrum could be fit equally well above 10 keV with a double power-law electron spectrum alone or with an isothermal component and a double power-law function above an electron energy of 18 keV. Also, it could be fitted with a multi-temperature thermal function over the full energy range. If they assumed that all of the emission above 10 keV was from a nonthermal distribution of high energy electrons, then they arrived at the unlikely conclusion that most of the flare energy was released prior to the major HXR burst. Thus, it seems likely that at least some of the emission may have been from plasma at temperatures as high as several tens of MK.

The separation of thermal and nonthermal emission can be aided by the analysis of the Fe-line features in the spectrum below 10 keV as discussed in Sect. 3.1, since it is safe to assume that the Fe-line emission is from thermal plasma. While it is possible that inner-shell lines ($K\alpha$, $K\beta$) can be generated by impact ionization with high-energy particles, no evidence for such a production process has ever been detected using the various high-resolution crystal spectrometers on SMM, Yohkoh, and Coronas-F (Phillips, private communication; see [14] for a possible counter-example). This, then, has consequences for the continuum that must accompany the line emission. The temperature, iron abundance, and line flux that can be obtained from the RHESSI observations of the iron-line feature can be used to constrain the possible thermal continuum, and, hence, allow the nonthermal component to be more accurately estimated in the region of overlap.

4.2 Hard X-ray Flares and Escaping Electrons

Some of the electrons accelerated in a flare lose their energy by collisions in the denser, lower solar atmosphere producing the HXR emission seen with RHESSI, while others escape into interplanetary space. Consequently, an interesting comparison can be made between RHESSI flare HXR measurements and the in situ electron measurements made in the vicinity of the Earth. Whether the HXR-producing electrons and the escaping electrons are accelerated by the same mechanism is not known. Combining RHESSI HXR observations with in situ observations of energetic electrons near 1 AU from the WIND spacecraft [35] allows a detailed temporal, spatial, and spectral study to be made for the first time. Early results show that events with a close temporal agreement between the HXR and the in-situ detected electrons (taking the time of flight of the escaping electrons into account) show a correlation between the HXR photon spectral index and the electron spectral index observed in-situ [34] indicating a common acceleration mechanism. Furthermore, the X-ray source structure of these events looks similar, showing hot loops with HXR footpoints plus an additional HXR source separated from the loop by ~15 arcseconds (Fig. 10 left). This source structure can be explained by a simple magnetic reconnection model with newly emerging flux tubes that reconnect with previously open field lines as shown in Fig. 10 (right).

4.3 Hard X-ray Footpoints

Solar HXR bremsstrahlung from energetic electrons accelerated in the impulsive phase of a flare is generally observed to be primarily from the footpoints of magnetic loops (see Fig. 8 for an example). The mechanism that accelerates

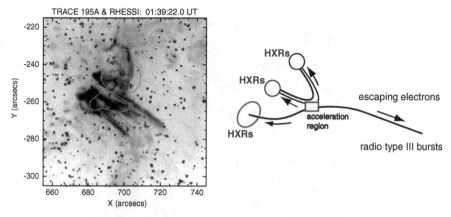

Fig. 10. EUV and X-ray sources of a flare that released energetic electrons into interplanetary space that were later observed near the Earth. **Left figure:** RHESSI contours at 6–12 keV (*red or dark gray: thermal emission*) and 20–50 keV (*blue or black: non-thermal emission*) overlaid on a TRACE 195Å EUV image (*dark region corresponds to enhanced emission*). Located at around [700, -245] arcsec, the X-ray emission outlines a loop with two presumably nonthermal footpoints. The strongest footpoint source however, is slightly to the southeast [683, -257] and shows a surprisingly lower intensity thermal source. **Right figure:** Suggested magnetic field configuration showing magnetic reconnection between open and closed field lines inside the red or dark gray box marked as the "acceleration region" where downward moving electrons produce the HXR sources and upward moving electrons escape into interplanetary space

the electrons is still not known but standard 2D magnetic reconnection models predict increasing separation of the footpoints during the flare (e.g., [51]) as longer and larger loops are produced. If the reconnection process results in accelerated electrons [48], the HXR footpoints should show this apparent motion. The motion is "apparent" because it is due to the HXR source shifting to footpoints of neighboring, newly reconnected field lines. Hence, the speed of footpoint separation reflects the rate of magnetic reconnection, and should be roughly proportional to the total HXR emission from the footpoints. Sakao, Kosugi, & Masuda [60] analyzed footpoint motions in 14 flares observed by Yohkoh HXT, but did not find a clear correlation between the footpoint separation speed and the HXR flux. Recently, however, [48] found some correlation between the source motion seen in Hα and the HXR flux during the main peak of a flare, but not before or after. Fletcher & Hudson [15] carried out similar analysis of footpoint motion using early RHESSI observations of several GOES M-class flares. They found systematic, but more complex footpoint motions than a simple flare model would predict.

Footpoint motion can be best studied in the largest events that show intense HXR emission lasting for many minutes. Krucker, Hurford, & Lin [33] analyzed HXR footpoint motions in the July 23, 2002 flare (GOES X4.8).

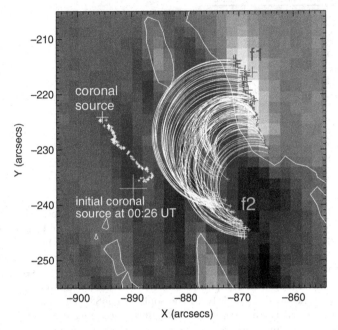

Fig. 11. Temporal evolution of the HXR sources seen with RHESSI during the flare on 23 July 2002. The motions of footpoints f1 (*black crosses*), f2 (*white crosses*), and f3 at 30–80 keV and the coronal source (*white crosses*) at 18–25 keV are indicated by the increasing size of the symbols to represent times from 00:26:35 to 00:39:07 UT. The centroid positions of the different sources are shown every 8 s for the footpoints and every 26 s for the coronal source. The semicircles connect simultaneously brightening footpoints. Note that the coronal source seems to be at a higher altitude than the tops of these ad-hoc semicircles. The grey-scale image is an MDI magnetogram in which the apparent neutral line is shown in white and the extreme line-of-sight values of the magnetic field are ± 600 G (from [33])

As can be seen in Fig. 11, at least three HXR sources above 30 keV can be identified during the impulsive phase with footpoints of coronal magnetic loops in an arcade. On the northern ribbon of this arcade, a source (f1) is seen that moves systematically along the ribbon for more than 10 minutes. On the southern ribbon, at least two sources (f2 and f3) are seen that do not seem to move systematically for longer than 30 s, with different sources dominating at different times. The northern source motions are fast during times of strong HXR flux, but almost absent during periods with low HXR emission. This is consistent with magnetic reconnection if a higher rate of reconnection of field lines (resulting in a higher footpoint speed) produces more energetic electrons per unit time and therefore more HXR emission. The absence of footpoint motion in one ribbon is inconsistent with simple reconnection models.

An additional correlation predicted from the simple theoretical reconnection model is between the footpoint motion and the rate of energy deposited by the energetic electrons into the footpoints. The idea is that the higher the reconnection rate, the more electrons are accelerated and the faster the footpoints move apart. The rate of energy deposition into the footpoints can be readily determined from the RHESSI imaging spectroscopy observations of the 23 July 2002 event assuming thick-target interactions. Combining this with the footpoint velocities derived from Fig. 11 gives a rough correlation as expected. Since this flare was close to the limb, the magnetic field strength could not be well determined in the footpoints, but assuming a constant value of 1000 G, we get the reconnection rates varying between $\sim$1 and 5×10^{18} Mx s^{-1}. These results are consistent with a model in which a higher rate of magnetic reconnection makes the footpoints move faster and also accelerates more electrons and deposits more energy at the footpoints.

5 Gamma Rays

The acceleration of ions to high energies in large solar flares has been established by the detection of nuclear gamma-ray line emission (e.g., [9]). When energetic ions collide with the solar atmosphere, they produce excited nuclei that emit prompt nuclear de-excitation lines, as well as secondary neutrons and positrons that result in the delayed 2.223 MeV neutron-capture line and the 511 keV positron-annihilation line [54]. Because of Doppler broadening, the line widths are dependent on the temperature or the velocity distribution of the emitting particles. High-energy protons interacting with the heavy ions of the ambient atmosphere produce relatively narrow gamma-ray lines characteristic of the different elements. High-energy heavier ions interacting with ambient protons, on the other hand, produce much broader lines because of the high velocities of the emitting ions. Also, accelerated α particles give detectable line features below 500 keV when they interact with ambient helium nuclei, the so-called α-α interactions. Spectral observations of all of these features, both the narrow and broad lines, provide information on the energy spectrum and composition of the accelerated ions and on the composition of the ambient target atmosphere (e.g., [8, 64]). All of these gamma-ray emissions are evident in the RHESSI spectrum shown in Fig. 12 for the X17 flare on 28 October 2003. The best-fit templates of the expected features are also shown for clarity. For the first time, RHESSI has the energy resolution necessary to resolve all of these gamma-ray lines, except for the intrinsically narrow 2.223 MeV line, and to determine the detailed line shapes expected from Doppler-shifts and the different possible velocity distributions. In addition, RHESSI provides the first spatial information on the gamma-ray sources, the only direct indication of the spatial properties of accelerated ions near the Sun.

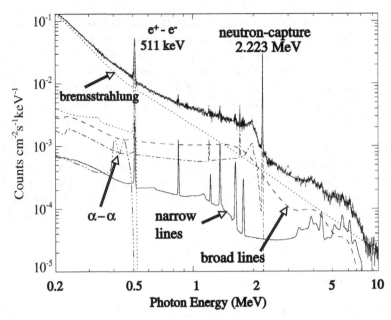

Fig. 12. RHESSI count-rate spectrum recorded during the X17 flare on 28 October 2003 from 11:06:20 to 11:10:04 UT. Clearly evident in this spectrum is the prominent neutron-capture line at 2.223 MeV, the positron annihilation line at 511 keV, numerous other narrow lines from accelerated protons, broad lines from accelerated heavy ions, and the features below 500 keV from α-α interactions. Underlying the line features is the bremsstrahlung continuum from accelerated electrons. Fitted templates of the expected spectra for the different components are also indicated. (After [67])

5.1 The 511 keV Positron-Annihilation Line

Positrons are produced in solar flares from the decay of both radioactive nuclei and pions that themselves are the result of interactions of the flare accelerated ions in the solar atmosphere. The production of observable gamma-rays from the positrons is not a straightforward process. A fortunate consequence of the complications is that much unique information can be obtained about the ambient medium through which the positrons pass from production to annihilation. Before the positrons can interact with the ambient thermal electrons, they must slow down by collisions until they have similar velocities. Then, a positron can either annihilate directly with a bound or a free electron to produce two 511-keV photons traveling in opposite directions, or it can combine with a bound or free electron to produce a hydrogen-like positronium 'atom' consisting of the positron and an electron in orbit around one another. After a while, the positronium decays in one of two ways depending on the relative spins of the two particles. If the spins are antiparallel (the singlet state), then the two particles annihilate, again with the production of two oppositely directed 511-keV photons. If the spins are parallel (the triplet state), then three photons are produced, all with energies below 511 keV. This latter case is observable as a continuum below the 511-keV line. The continuum-to-line ratio is called the $3\gamma/2\gamma$ ratio and its determination is an important goal of the spectral analysis of RHESSI data for individual flares.

Further complications arise because the positronium can be formed either by thermal charge-exchange when essentially at rest, or by charge-exchange in flight. In the first case, the subsequent decay results in a very narrow 511-keV line ($\sim$1.5 keV FWHM). In the second case, the resulting 511-keV line can be Doppler broadened to $\sim$7.5 keV FWHM, easily measurable with RHESSI. At densities above $\sim 10^{13}$ cm^{-3}, the 3γ continuum intensity is reduced as a result of collisions that cause transitions from the triplet to the singlet state and the breakup of the positronium, thus affecting the $3\gamma/2\gamma$ ratio.

Although the positron story is complicated, all the processes are well understood. Through careful modeling, the measured 511-keV line shape, the $3\gamma/2\gamma$ ratio, and the time history can provide information on the temperature, density, composition, and ionization state of the ambient medium in which the positrons slow down, form positronium, and annihilate. In this sense, then, the positrons act as a thermometer, a barometer, and an ionization gauge for the ambient solar atmosphere.

RHESSI has now observed at least four flares with sufficient intensity to provide useful diagnostics from the positron-electron annihilation line. The first flare with high-resolution spectral observations of this line occurred on 2002 July 23 [65]. In that case, the line had a Gaussian width of 8.1 ± 1.1 keV (FWHM). Two interpretations of this width are possible: either the annihilations took place in a medium with temperatures as high as $\sim (4-7) \times 10^5$ K or the positronium formation took place by charge exchange in flight at temperatures near 6000 K.

Better statistics were obtained for the X17 flare that started at 09:51 UT on 28 October 2003 [66]. The line profile is shown in Fig. 13 for two different times during the flare. Early in the flare, the annihilation line was broadened to 8 ± 1 keV(FWHM), suggesting temperatures of $> 2 \times 10^5$ K if thermal. Later during the flare, at a time when observations at other wavelengths show that nothing unusual appears to be happening, the line becomes much narrower ($\sim$1 keV) on time scales of a few minutes, suggesting temperatures of less than or $\sim 1 \times 10^4$ K and high ionization levels. In fact, the line became the narrowest that the RHESSI spectrometer has ever measured, in space or on the ground. In Fig. 13, the change in the width of the 511-keV line is evident suggesting a dramatic reduction in temperature of the annihilation region. However, this sudden narrowing poses serious problems for a thermal interpretation. What heats the chromospheric material with densities between 10^{12} and 10^{14} cm^{-3} to such high temperatures and why does it cool so suddenly with no other obvious manifestation?

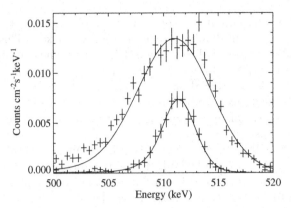

Fig. 13. Count spectra of the 511-keV positron-annihilation line, after subtracting bremsstrahlung, nuclear, and induced 511-keV line contributions, for two different times during the 28 October, 2003, X17 flare. The broad line was obtained between 11:06 and 11:16 UT, and the narrow line between 11:18 and 11:30 UT. The *solid curve* in each case is the best-fitting model that includes a Gaussian line centered at 511-keV and the positronium continuum at lower energies. (After [66])

5.2 Nuclear De-excitation Lines

RHESSI has obtained the first high-resolution measurements of nuclear de-excitation lines produced by energetic ions accelerated in solar flares [71]. Narrow lines from high-energy proton interactions with ambient neon, magnesium, silicon, iron, carbon, and oxygen, resolved for the first time, are shown in Fig. 14 for the flare of 2002 July 23 at a heliocentric angle of ∼73°. The deviation of the lines from their rest-frame energy and the measured line widths indicate Doppler redshifts of 0.1–0.8% and line broadening of 0.1–2.1% (FWHM). These values generally decrease with the atomic mass of the emitting nucleus, as expected, since heavier nuclei will recoil less from a collision with a fast proton or α particle. The measured redshifts for this flare are larger than expected for a model of an interacting ion distribution isotropic in the downward hemisphere in a radial magnetic field. To explain these observations, either the ions are traveling along a magnetic loop inclined towards the Earth at ∼40° to the radial direction, or the ions are highly beamed. Bulk downward motion of the plasma in which the accelerated ions interact can be ruled out [71].

5.3 The 2.223 MeV Neutron-capture Line

Spectroscopy

The production of the intrinsically narrow gamma-ray line at 2.223 MeV, while not as complicated as that of the positron annihilation line, is also subject to various effects that make analysis difficult but potentially very informative.

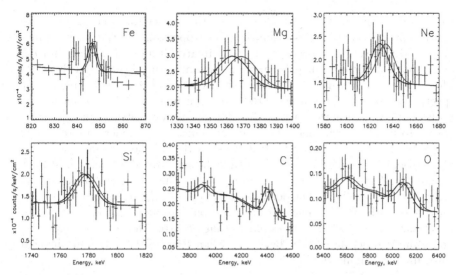

Fig. 14. RHESSI background-subtracted count spectra from 00:27:20 to 00:43:20 UT on 2002 July 23. Each panel is labeled with the element primarily responsible for the line shown. The carbon and oxygen lines also show the secondary peak from the escape of a 511 keV positron-annihilation photon, which also contains information on the line shape. The *thick curve* shown in each panel is the Gaussian fit from Table 1 plus the underlying bremsstrahlung continuum and broad lines (see text), convolved with the instrument response. The thinner line is the same fit forced to zero redshift for comparison. The error bars are $\pm 1\sigma$ from Poisson statistics. (From [71])

High-energy neutrons are the products of interactions of the accelerated ions with the ambient solar atmosphere. Before they can be captured by ambient thermal hydrogen atoms, however, they must have similar velocities. Consequently, there is a delay while they lose their energy through collisions. Since the neutrons are not constrained by the magnetic fields and can travel relatively long distances, they can penetrate down to photospheric levels before becoming thermalized. After a minute or two, they are captured by hydrogen to form deuterium with the immediate release of a 2.223 MeV gamma-ray carrying the excess binding energy. The line is intrinsically only ~0.1 keV wide since the deuterium is essentially at rest when it is formed.

The first flare observed by RHESSI to show 2.223 MeV line emission occurred on 23 July 2002. The measured FWHM line width was ~4 keV, as expected from the germanium spectral resolution at that energy. The intensity of the line was such that its time history could be determined for a period of about 20 minutes with integration times of 20 s. Murphy et al. [46] have compared the measured time histories with the predictions of a comprehensive model for particle transport in a magnetic loop that includes Coulomb collisional losses, magnetic mirroring, and pitch-angle scattering. The effects of neutron capture by ^{3}He and the angular distribution of the accelerated ions

that produce the neutrons are also factors that affect the time history. The
neutron production rate was assumed to be proportional to the observed flux
of 4- to -7.6 MeV gamma-rays produced primarily from nuclear de-excitation
of carbon and oxygen. The predicted and measured time histories of the 2.223
MeV line are shown in Fig. 15. Here, the free parameters are the power-law
spectral index (s) of the accelerated ions (taken to be 4.5), the mean free
path (λ) to isotropize the particle distribution through pitch-angle scattering
(taken to be $\lambda = 2000$ times the loop half length, i.e., moderate pitch-angle
scattering), and the ^{3}He/H ratio (taken as 7×10^{-5}). The agreement with the
observations is remarkable and shows that the accelerated particles must have
suffered moderate pitch-angle scattering during their transport through the
coronal part of the loop. The derived ^{3}He/H ratio could be better constrained
for a flare with stronger nuclear de-excitation line fluxes. Similar analysis is
being carried out for the intense flares that occurred during the three week
period in October and November 2003, when RHESSI observed and imaged
three further gamma-ray line flares.

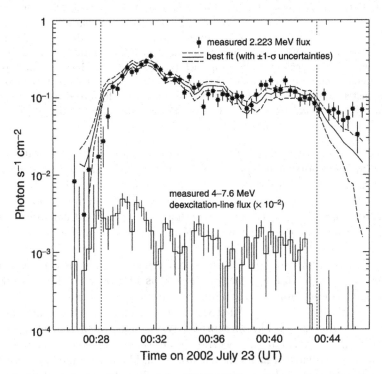

Fig. 15. Measured time history of the 2.223 MeV neutron-capture line compared
with the best-fitting predicted line fluxes for s = 4.5 (obtained with $\lambda = 2000$ and
^{3}He/H = 7×10^{-5}). The *dotted lines* indicate the time interval over which the model
was calculated (00:28:20 to 00:43:20 UT). The 4-to-7.6 MeV de-excitation line fluxes
are also shown, reduced by a factor of 100 for clarity. (After [46])

Imaging

The highest sensitivity for gamma-ray imaging can be achieved by using the 2.223 MeV neutron-capture line. This is because of relatively good statistics and the intrinsically narrow line width, which minimizes the bremsstrahlung continuum contribution and the non-solar background that must be included compared to the broader lines. Hurford et al. [27] reported a single source structure at this energy for the flare on July 23, 2002, the first gamma-ray-line flare observed by RHESSI. The source was unresolved at the instrumental resolution of 35"; a diffuse source of greater extent than this was excluded by the observations. Surprisingly, the 2.2-MeV source centroid was displaced by 20±6" from the centroid of the HXR sources. The series of very large flares occurring in October/November 2003 confirmed this finding [28]. For the event with the best statistics, on October 28, 2003, the 2.2 MeV image in Fig. 16 shows two sources similar to the HXR footpoint sources separated by ∼70

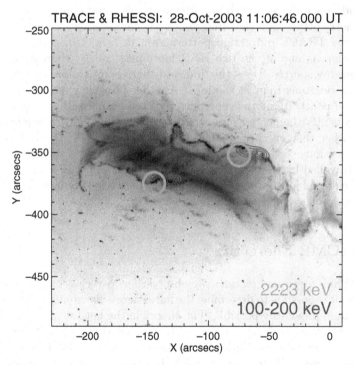

Fig. 16. Imaging of the 2.223 MeV neutron-capture line and the HXR electron bremsstrahlung of the flare on October 28, 2003. The *red or gray circles* show the locations of the event-averaged centroid positions of the 2.223 MeV emission with 1σ uncertainties; the *blue or black lines* are the 30, 50, and 90% contours of the 100–200 keV electron bremsstrahlung sources at around 11:06:46UT. The underlying EUV image is from TRACE at 195Å with offset corrections applied. The gamma-ray and HXR sources are all located on the EUV flare ribbons seen with TRACE

arcsec. However, both the gamma-ray sources appear to be displaced from the corresponding HXR source by ~15".

The displacement of the gamma-ray and HXR sources suggests that energetic electrons and ions lose their energy at different places in the solar atmosphere. This could be because the electrons and ions are accelerated at different locations or because of different transport effects from a possibly common acceleration site. Emslie et al. [17] noted that a stochastic acceleration model based on cascading MHD turbulence proposed by [44] predicts that ion acceleration takes place in the vicinity of large loops or where the Alfven speed is low, while the electron acceleration originates in shorter loops or where the Alfven speed is high. Ion acceleration in long loops and electron acceleration in short loops would explain the different source locations seen in the 23 July flare. However, the October 28, 2003 observations showed similar separations between the two sources seen in HXRs and between the two seen in the 2.2 MeV emission (Fig. 16) suggesting acceleration in similar sized loops. The alternate explanation that electrons are accelerated in regions with high Alfven speeds and ions in regions with low Alfven speeds could apply in both cases.

It is noteworthy that the brightest EUV emission, as indicated by the origin of the TRACE diffraction pattern seen in (Fig. 16), comes from the same location as one of the two HXR footpoints, rather than from one of the gamma-ray sources. Does this indicate that there is more energy in the accelerated electrons than in the ions? Another possible explanation is that the ions will penetrate more deeply into the chromosphere than the electrons and consequently the heated plasma will be cooler and emit preferentially in the UV rather than in the TRACE 195Å passband. It is unlikely that the different locations of the X-ray and gamma-ray sources is the result of the distance travelled by the neutrons from their point of origin where the bulk of the accelerated ions interact to the place where they become thermalized; this separation is estimated to be only ~500 km.

6 Flare/CME Energetics

With RHESSI X-ray and gamma-ray observations covering such a broad energy range, it is possible to determine the flare energy distribution with greater precision than previously possible. The energy in the hottest plasmas, the accelerated electrons, and the ions can all be estimated from RHESSI observations with better than the order-of-magnitude accuracy that has previously been possible. Saint-Hilaire and Benz [56] obtained an energy budget for a flare on 26 February 2002 using RHESSI and TRACE observations. They found that the energy in the nonthermal electrons producing the HXRs was more than an order of magnitude greater than the thermal and radiated energy in the flare kernel plus the kinetic energy of the jet seen with TRACE. This rather surprising result is consistent with similar conclusions reported by [11] based on earlier SMM observations and analysis of the RHESSI observations of the 20 February 2002 flare shown in Figs. 8 and 9. Saint-Hilaire and Benz [57] have recently reported on the analysis of RHESSI observations for 9 medium-sized flares, and find that the thermal and nonthermal energies are of the same magnitude.

By combining the RHESSI spectral results with flare observations at other wavelengths, a differential emission measure analysis is being pursued using the Markov-Chain Monte Carlo (MCMC) method described by [32]. This will allow the energetic contributions of the lower temperature gas to be estimated and hence, to determine if it provides a significant contribution to the thermal energy of a flare.

Emslie et al. [17] report on the first attempt to obtain comprehensive energy budgets for two X-class flares and the associated CMEs, one on 21 April 2002 and the second on 23 July 2002. In addition to the RHESSI flare data, the energetics of the associated CMEs were determined from the SOHO/LASCO observations, and the energy in high-energy particles in space was estimated from the in situ measurements with instruments on ACE, SOHO, and Wind. Unfortunately, the uncertainties on all the different components of the energy are large. The major limitations are that the quoted energies obtained from the RHESSI X-ray and gamma-ray observations are all lower limits. Although reducing the filling factor from the assumed value of $f = 1$ to a possible value as low as 10^{-4} reduces the energy estimate by $\sqrt{f}$, the thermal energy is probably a lower limit because of the underestimate of the radiative and conductive cooling losses. The energy in electrons is a lower limit because of uncertainties in the low-energy cutoff to the electron spectrum. The energy in accelerated ions is a lower limit for the 2002 July 23 flare because of the unknown contribution from ions below a few MeV. (No gamma-ray lines were seen during the 2002 April 21 flare.)

Despite these limitations, [17] drew the following tentative conclusions from the results:

1. For the 23 July 2002 flare, the energy in accelerated electrons is comparable to the energy in accelerated ions, in agreement with the conclusion reached by [55] based on SMM results for 19 flares.
2. The CME energy dominates over the combined flare energies in both events.
3. The CME energy is a substantial fraction ($\sim$30%) of the available magnetic energy in both events.

However, conclusion (2) was based on the assumption that the radiant energy from each of the two flares under study was only a factor of two greater than the peak energy in the thermal X-ray emitting plasma. This is contrary to the factor of $\sim$5–20 obtained from the estimate based on the relationship between the soft X-ray ($L_{softX-rays}$) and total (L_{total}) luminosities given by [24] and corrected by [69], i.e. $L_{softX-rays}/L_{total} = 2/30$. This relationship for the 21 April 2002 flare gives a total radiant energy of $10^{31.7}$ ergs. This must be compared to the Emslie et al. values for the available magnetic energy of

$10^{32.3 \pm 0.3}$ ergs, the peak energy in the thermal plasma of $10^{31.1}$ ergs, and radiative losses of $10^{31.3}$ ergs. If the higher value of the radiant energy is accepted, then the flare and CME energies are comparable both to one another and to the available magnetic energy. Clearly, given the order-of-magnitude uncertainties in the flare energies, the energetic dominance of the flare or the CME has not been established by the current estimates for these events. The recent measurement of the total luminosity of the X28 flare on 2003 November 5 [79] should provide an accurate normalization of the $L_{softX-rays}/L_{total}$ ratio for that event.

7 Conclusions

We have tried to provide a representative sampling of the science results from RHESSI solar X-ray and gamma-ray observations. We have not attempted to summarize all of the over 180 RHESSI-related papers that have already been published according to the compilation maintained by Aschwanden at http://www.lmsal.com/~aschwand/publications/hessi.html.

Many of the early results were as expected based on previous observations but some have been particularly surprising. Perhaps the most surprising is the apparent displacement between the source of the neutron-capture gamma-ray line and the source of bremsstrahlung X-rays suggesting differences in the acceleration and/or transport of the energetic ions and electrons. The measured gamma-ray line redshifts were expected but their higher than predicted magnitude suggests that the energetic ions were highly beamed or that they traveled along a highly inclined magnetic loop. The intensity of the 511-keV positron annihilation line shows that ~1 kg of antimatter was produced in the 23 July flare, but how can we interpret the initial heating of the target chromospheric material to such high temperatures that is apparently required by the measurements of the line width. Perhaps a more difficult question is, how does the target material cool so suddenly?

In the hard X-ray domain, RHESSI's high energy resolution has allowed for the most convincing separation to date of the thermal and nonthermal components of the emission. In some cases, the nonthermal spectrum extends down to as low as 10 keV, thus increasing our estimates of the nonthermal energy based on a 25-keV cutoff energy by an order of magnitude or more for average events and by a factor of ~500 for microflares. The flare energy budget calculations have been aided by RHESSI observations in this way but the energy estimates for the different components are still bedeviled by unknown filling factors and cooling terms. At higher HXR energies, the interest has been on the downward break in the hard X-ray spectrum often seen between ~20 and ~100 keV, suggesting that the electron acceleration process must produce a corresponding break in the electron spectrum. The significance of this for acceleration models has not yet been fully explored.

In its lowest energy domain, RHESSI has provided many interesting observations, thanks to its great sensitivity and sufficient energy resolution to measure the very steep spectra and detect the Fe-line features, even in the more intense flares. The most remarkable result is perhaps the evidence of magnetic reconnection in a current sheet above the flare loops based on the observation of an above-the-looptop source that had a temperature gradient with altitude opposite to that of the underlying loop source. The initial apparent downward motion of this coronal source observed in several flares was a surprise.

Clearly, RHESSI observations are fulfilling the objectives of helping to understand energy release and particle acceleration in solar flares. Much remains to be done in analyzing the many flare observations already made. The instrument is still fully operational and continues to make new flare observations, even as we move towards solar minimum. RHESSI has no consumables, meaning that, barring some unforseen failure of a critical function, it can continue to operate until it re-enters the atmosphere. This will not be, at the earliest, until the rise in solar activity again heats and expands the outer atmosphere. Thus, it is hoped that RHESSI can be kept operating well into the next maximum of activity, allowing it to continue making unique observations of flares over a complete solar cycle and detecting more of the very rare gamma-ray-line events that are so revealing.

Acknowledgments

RHESSI is named for the late Reuven Ramaty, a co-investigator at Goddard Space Flight Center and a pioneer in the fields of solar physics, gamma-ray astronomy, nuclear astrophysics, and cosmic rays.

We thank the whole RHESSI team - without their dedicated effort, none of this would have been possible. We are also indebted to the innumerable scientists around the world who have participated in the analysis and scientific interpretation of the observations. Thanks for sharing your work so freely and for participating in the numerous workshops that have contributed so wonderfully to the cross-fertilization of information and ideas.

For reviewing early drafts of the paper, we thank Leah Haga, Gordon Holman, Gordon Hurford, Ron Murphy, Kenneth Phillips, Gerry Share, and Linhui Sui. We thank Amir Caspi for Fig. 3, Gerry Share for Fig. 12, and Gordon Hurford for Fig. 16, all of which were made available to us prior to publication. We thank the referee, Gordon Emslie, for doing such a careful job that significantly improved the accuracy and readability of the paper.

Finally, we thank Ludwig Klein for inviting us to the CESRA Workshop on the Isle of Skye in Scotland and for insisting so strenuously and persistently that we write this paper.

HSH and SK are supported in part by NASA contract NAS5-98033.

References

1. Alexander, R. C. and Brown, J. C. 2002, Solar Physics, 210, 407–418.
2. Aschwanden, M. J., Brown, J. C., and Kontar, E. P. 2002, Solar Physics, 210, 383–405.
3. Bai, T. and Ramaty, R. 1978, ApJ., 219, 705.
4. Benz, A. O., Grigis, P. C. 2002, Sol. Phys., 210, 431.
5. Brown, J. C., van Beek, H. F., and McClymont, A. N. 1975, A&A, 41, 395–402.
6. Brown, J. C. 1973, Solar Physics, 28, 151.
7. Caspi, A., Krucker, S., and Lin, R. P. 2004, COSPAR presentation, COSPAR04-A-03582 E2.3-0013-04.
8. Chupp, E. L. 1984, ARA&A, 22, 359.
9. Chupp, E. L. 1990, Phys. Scr., T18, 15.
10. Crosby, N. B., Aschwanden, M. J., Dennis, B. R. 1993, Sol. Phys., 143, 275.
11. Dennis, B. R., Veronig, A., Schwartz, R. A., Sui, L., Tolbert, A. K., and Zarro, D. M. 2003, Advances in Space Research, v. 32, no. 12, pp. 2459–2464.
12. Dennis, B. R., Phillips, K. J. H., Sylwester, J., Sylwester, B., Schwartz, R. A., and Tolbert, A. K., Advances in Space Research, 2005, 35(10), 1723–1727.
13. Dere, K. P., Landi, E., Mason, H. E., Monsignori Fossi, B. C., & Young, P. R., CHIANTI – An Atomic Database for Emission Lines, 1997, A&AS, 125, 149.
14. Emslie, A. G., Phillips, K. J. H., and Dennis, B. R. 1986, Slar Physics, 103, 89–102.
15. Emslie, A. G., Kontar, E. P., Krucker, S, Lin, R. P. 2003, ApJL, 595, L107.
16. Emslie, A. G. 2003, ApJL, 595, L119–L121.
17. Emslie, A. G., Kucharek, H., Dennis, B. R., Gopalswamy, N., Holman, G. D., Share, G. H., Voulidas, A., Forbes, T. G., Gallagher, P. T., Mason, G. M., Metcalf, T. R., Mewaldt, R. A., Murphy, R. J., Schwartz, R. A., and Zurbuchen, T. H. 2004, J. Geophys. Res., 109, A10104.
18. Emslie, A. G., Miller, J. A., Brown, J. C., 2004 ApJL, 602, L69.
19. Ewell, M. W. (jr.), Zirin, H., Jenson, J. B., and Bastian, T. S. 1993, ApJ, 403, 426–433.
20. Feldman, U. 2005, in preparation.
21. Fletcher, L., Hudson, H.S. 2002, Solar Physics, 210, 307.
22. Holman, G. D., Sui, L., Schwartz, R. A., and Emslie, A. G. 2003, ApJL, 595, L97–L101.
23. Hudson, H. S. 2000, ApJ, 531, L75–L77.
24. Hudson, H. S. 1991, Solar Physics, 133, 357–369.
25. Hudson, H. S., and Khan, J. I., 1996 in: Magnetic Reconnection in the Solar Atmosphere, ed. by R. D. Bentley and J. D. Mariska (ASP, San Francisco 1996), ASP Conf. Ser. 111, 135.
26. Hurford, G. J., Schmahl, E. J., Schwartz, R. A., et al., 2002, Solar Physics, 210, 81–86.
27. Hurford, G. J., Schwartz, R. A., Krucker, S., Lin, R. P., Smith, D. M., Vilmer, N. 2003, ApJL, 595, L77.
28. Hurford, G. H., Krucker, S., and Lin, R. P., 2006, ApJL, 644, L93–L96.
29. Kontar, E. P., Brown, J. C., and McArthur, G. K. 2002, Solar Physics, 210, 419–429.
30. Kontar, E. P., Emslie, A. E., Piana, M., Massone, A. M., Brown, J. C. 2005, Solar Physics, 226, 317–325.

31. Krucker, S., Christe, S., Lin, R. P., Hurford, G. J., Schwartz, R. A. 2002, Sol. Phys., 210, 445.

32. Krucker, S. and Lin, R. P. 2002, Solar Physics, 210, 229–243.

33. Krucker, S., Hurford, G. J., Lin, R. P. 2003, ApJL, 595, L103.

34. Krucker, S., Kontar, E., Lin, R. P. 2006, ApJ, submitted.

35. Lin, R. P. et al., 1995, Space Sci. Rev., 71, 125.

36. Lin, R. P. et al., 2002, Solar Physics, 210, 3–32.

37. Lin, R. P., Krucker, S., Holman, G. D., Sui, L., Hurford, G., and Schwartz, R. A. 2003, 28th international Cosmic Ray Conf., pp. 3207–3210.

38. Lin, L., Kashyap, V. L., Drake, J. J., DeLuca, E. E., Weber, M., and Sette, A. L. 2004, Proc. 13th Cool Stars Workshop, Hamburg, eds. F. Favata et al.

39. Liu, C., Qiu, J. Gary, D. E., Krucker, S., Wang, H. 2004, ApJ 604, 442.

40. Masuda, S., Kosugi, T., Hara, H., Tsuneta, S., Ogawara, Y. 1994, Nature, 371, 495.

41. Mazzotta, P., Mazzitelli, G., Colafrancesco, S., & Vittorio, N., Ionization Balance for Optically Thin Plasmas, 1998, A&AS, 133, 403.

42. McKenzie, D. E., Hudson, H. S. 1999, ApJ, 519, L93.

43. Metcalf, T. R., Alexander, D., Hudson, H. S., and Longcope, D. W. 2003, ApJ, 595, 483–492.

44. Miller, J. A. In: High Energy Solar Physics: Anticipating HESSI, ed by R. Ramaty & N. Mandzhavidze (ASP, San Francisco 2000), ASP Conf. Ser. 206, 145.

45. Morita, S., Uchida, Y., Hirose, S, Uemura, S., and Yamaguchi, T. 2001, Solar Physics, 200, 137–156.

46. Murphy, R. J., Share, G. H., Hua, X.-M., et al., 2003, ApJL, 595, L93–L96.

47. Ohyama, M., & Shibata, K. 1998, ApJ, 499, 934.

48. Øieroset, M., Lin, R. P., Phan, T. D., Larson, D. E., and Bale, S. D. 2002, Physical Review Letters, 89, 19.

49. Phillips, K. J. H. 2004, ApJ, 605, 921–930.

50. Piana, M., Massone, A. M., Kontar, E. P., Emslie, A. G., Brown, J. C., and Schwartz, R. A., ApJL, 595, L127–L130.

51. Priest, E. R. & Forbes, T. G. 2002, Astronomy Astrophysics Revs. 10, 313.

52. Qiu, J., Lee, J., Gary, D. E., Wang, H. 2002, ApJ, 565, 1335.

53. Qiu, J., Liu, C., Gary D. E., Nita, G. M, Wang, H. 2004, ApJ, 612, 530–545.

54. Ramaty, R., & Murphy, R. J. 1987, Space Sci. Rev., 45, 213.

55. Ramaty, R., Mandzhavidze, Kozlovsky, B., and Murphy, R. J. 1995, ApJL, 455, L193–L196.

56. Saint-Hilaire, P. and Benz, A. O. 2002, Solar Physics, 210, 287–306.

57. Saint-Hilaire, P. and Benz, A. O. 2005, Astron. Astrophys., 435, 743–752.

58. Sakao, T. 1994, PhD Thesis, University of Tokyo.

59. Sakao, T., Kosugi, T., Masuda, S., Yaji, K., Inda-Koide, M., and Makishima, K. 1995, Advances in Space Research, v. 17, no. 4/5, pp. 67–70.

60. Sakao, T., Kosugi, T., Masuda, S. 1998, in ASSL Vol. 229: Observational Plasma Astrophysics.

61. Schmahl, E. J. and Hurford, G. J. 2002a, Solar Physics, 210, 273–286.

62. Schmahl, E. J. and Hurford, G. J. 2002b, Adv. Space Res., 32, 2477–2482.

63. Schwartz, R. A. et al., 2002, Solar Physics, 210, 165.

64. Share, G. H., & Murphy, R. J. 1995, ApJ, 452, 933.

65. Share, G. H., et al., 2003, ApJL, 595, L85–L88.

66. Share, G. H., Murphy, R. J., Smith, D. M., Richard, A., Schwartz, R. A., and Lin, R. P. 2004, ApJL, 615, L169–L172.
67. Share, G. H. et al., 2005, in preparation.
68. Shibata, K., Masuda, S., Shimojo, M., Hara, H., Yokoyama, T., Tsuneta, S., Kosugi, T., Ogawara, Y. 1995, ApJ, 451, L83.
69. Shimizu, T. 1995, PASJ, 47, 251–263.
70. Smith, D. M., Lin, R. P., Turin, P., et al., 2002, Solar Physics, 210, 33–60.
71. Smith, D. M., Share, G. H., Murphy, R.J., Schwartz, R. A., Shih, A. Y., Lin, R. P. 2003, ApJL, 595, L81–L84.
72. Sui, L., Holman, G. D. 2003, ApJL, 596, L251.
73. Sui, L., Holman, G. D., and Dennis, B. R. 2004, ApJ, 612, 546–556.
74. Sui, L., Holman, G. D., and Dennis, B. R., Krucker, S., Schwartz, R. A., and Tolbert, K. 2002, Solar Physics, 210, 245–259.
75. Sylwester, J., Gaicki, I., Kordylewski Z., et al., RESIK: A Bent Crystal X-ray Spectrometer for Studies of Solar Coronal Plasma Composition, 2005, Solar Physics, 226, 45–72.
76. Tsuneta, S., Hara, H., Shimizu, T., Acton, L. W., Strong, K. T., Hudson, H. S., Ogawara, Y. 1992, PASJ, 44, L63.
77. Tsuneta, S. 1996, ApJ, 456, 840.
78. Veronig, A. M. & Brown, J. C. 2004, ApJL, 603, L117.
79. Woods, T. N., Eparvier, F. P., Fontenla, J., Harder, J., Kopp, G., McClintock, W. E., Rottman, G., Smiley, B., and Snow, M. 2004, Geophys. Res. Lett. 31, L10802.
80. Yokoyama, T., Akita, K., Morimoto, T., Inoue, K., Newmark, J. 2001, ApJ, 546, L69.

RHESSI Results – Time for a Rethink?

J. C. Brown[1], E. P. Kontar[1], and A. M. Veronig[2]

[1] Department of Physics and Astronomy, Kelvin Building, University of Glasgow, Glasgow G12 8QQ, Scotland, UK
john@astro.gla.ac.uk,eduard@astro.gla.ac.uk
[2] Institute for Geophysics, Astrophysics and Meteorology, University of Graz, Universitätsplatz 5, A-8010 Graz, Austria
asv@igam.uni-graz.at

Abstract. Hard X-rays and γ-rays are the most direct signatures of energetic electrons and ions in the sun's atmosphere which is optically thin at these energies and their radiation involves no coherent processes. Being collisional they are complementary to gyro-radiation in probing atmospheric density as opposed to magnetic field and the electrons are primarily 10–100 keV in energy, complementing the ($>$100 keV) electrons likely responsible for microwave bursts.

The pioneering results of the Ramaty High Energy Solar Spectroscopic Imager (RHESSI) are raising the first new major questions concerning solar energetic particles in many years. Some highlights of these results are discussed – primarily around RHESSI topics on which the authors have had direct research involvement – particularly when they are raising the need for re-thinking of entrenched ideas. Results and issues are broadly divided into discoveries in the spatial, temporal and spectral domains, with the main emphasis on flare hard X-rays/fast electrons but touching also on γ-rays/ions, non-flare emissions, and the relationship to radio bursts.

1 Introduction

Major observational results from RHESSI and instrumental details have been extensively described elsewhere (e.g. [1] and other articles in that volume, and [2]) and will not be repeated here. Based on results from numerous earlier spacecraft from OGOs, OSOs and TD1A through SMM, Hinotori and Yohkoh (these three giving the first HXR images), the conventional wisdom prior to RHESSI envisaged electron and ion acceleration high in a loop near a reconnection site. Most of the hard X-rays (HXRs) and γ-rays were believed to originate in two bright loop footpoints by collisional thick target deceleration of fast particles with a near power-law spectrum in the dense chromosphere [3], plus occasional fainter emission at or above the looptop as seen in Yohkoh [4] and sometimes even higher as seen in limb occulted flares [5]. Until RHESSI, apart from one balloon flight [6], spectral resolution was very limited, particularly in images and in (non-imaged) γ-rays. RHESSI has

J. C. Brown et al.: *RHESSI Results – Time for a Rethink?*, Lect. Notes Phys. **725**, 65–80 (2007)
DOI 10.1007/978-3-540-71570-2_4

transformed this via Ge detector spectrometry, yielding high resolution spectra and spectral images in HXRs, high resolution γ-ray line spectroscopy, and the first γ-ray line images. RHESSI also excels in having an unsaturated spectral range from a few keV to ten of MeV, thus yielding data on the hot SXR plasma as well as on fast particles (see articles in special issues of Solar Phys. Vol. 210, 2002 and Astrophysical Journal Letters Vol. 595, 2003). While many of the RHESSI data show events with some resemblance to the canonical thick target footpoint scenario, with near power-law spectra, there are many examples deviating from this simple picture. Here the main emphasis is on these new features as they are the driving force behind the need for a rethink.

2 Imaging Discoveries and Issues

Probably the most exciting imaging discovery by RHESSI is the fact that, in at least one of the few strong γ-ray line events seen by RHESSI, the 2.2 MeV neutron capture line comes from a spatial location quite distinct from the source of HXRs – Fig. 1 [7]. Cross-field transport is unable to explain this spatial separation and it seems it must be due to acceleration of electrons and of ions in or into quite distinct magnetic loops. The only explanation offered to date is that by Emslie, Miller and Brown [8] where the longer/shorter Alfvén travel time in larger/smaller loops favours respectively the stochastic

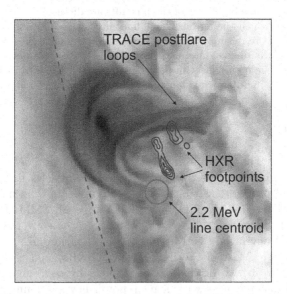

Fig. 1. Hard X-ray (electron) emission versus 2.2 MeV (nuclear) gamma-ray line emission centroid location for July 23, 2002 with TRACE context, based on [7]. The displacement of the fast ions from the fast electrons was one of the biggest surprises in RHESSI data

acceleration of ions/electrons. The Doppler profiles of the γ-ray lines also help constrain the geometry of the loop in which the same ions move.

Those events which do show a classic 2-footpoint structure (e.g. Fig. 2 (left)), at least within the spatial resolution limits of RHESSI, can in principle be compared with the predictions of the thick target model in terms of the spectral variation of the footpoint structure. Qualitatively, the highest energy electrons penetrate deepest so that the hardest HXR footpoints should lie lowest, and furthest apart in the loop, to an extent depending on the variation of electron energy loss rate with electron energy. On the conventional assumption of collision-dominated energy losses, Brown, Aschwanden

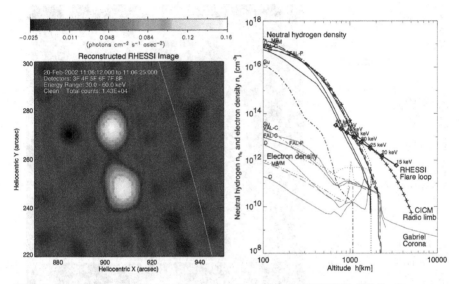

Fig. 2. Left: Simple 'classic' 2-footpoint flare seen by RHESSI. While there may be other interpretations, such a structure is expected if the field approximates to a simple bipolar loop without strong field convergence. Then electrons accelerated anywhere in the upper, low-density, loop can reach the dense chromosphere at the loop ends ('footpoints') where they emit bremsstrahlung very strongly compared with in the tenuous corona. Such high 'footpoint contrast' was discussed as early as Brown and McClymont [9] and MacKinnon, Brown and Hayward [10]. **Right**: Height distribution of hydrogen and other densities, as labelled, in numerous solar atmospheric models with superposed that required for a collisional thick target to match the RHESSI data for the flare of 2002, February 20 from [12]. The basis of [12] is to assume that 'footpoint flares' like that in Fig. 2 confirm the collisional thick target model of injection of electrons from the corona down the legs of a loop where they undergo purely collisional transport as they radiate. Since high energy electrons penetrate deeper, the footpoint centroid height should decrease with increasing energy and at a rate depending on the plasma density there. For an assumed electron injection spectrum one can then use the energy dependence of the HXR centroid height to infer the density as a function of height, the beam acting as a probe of the target

and Kontar [11] and Aschwanden, Brown and Kontar [12], determined the atmospheric density structure $n(h)$ needed for the thick target model to produce the observed spectral image structure, with the results shown for one event in Fig. 2 (right; February 20, 2002; for a HXR map see Fig. 2 (left)). These show that collisions are a substantial factor in electron transport and

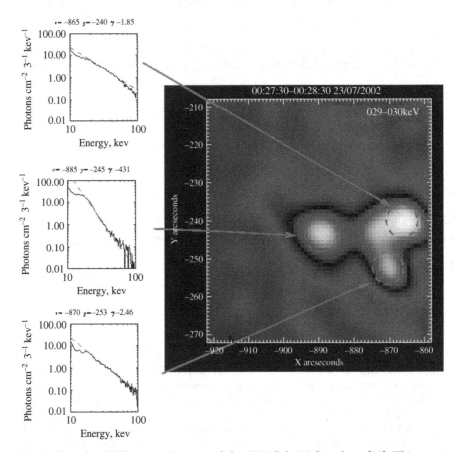

Fig. 3. Complex HXR spectral image of the 2002 July 23 flare from [13]. This event seems to show four distinct sources and does not conform to the simple bipolar pattern of Fig. 2. If, for example, the two rightmost patches were footpoints of a single loop and the leftmost one the looptop, then their local spectral indices – measurable accurately for the first time by RHESSI spectrometric imaging – are inter-related roughly as expected in the collisional thick target model, the footpoint spectra being roughly 2 powers harder than the looptop. But the looptop source is higher than expected and the fourth source is hard to explain in any simple way – cf. [13]. RHESSI's spectral resolution, sensitivity, and large dynamic range are enabling such questions to be asked for the first time. The evolution of the different RHESSI sources superposed on an MDI magnetogram is shown in Fig. 11 of the article by Dennis et al. [2] in this volume

may be the dominant factor if flare $n(h)$ is similar to spicules (Fig. 2, right). A lower $n(h)$ structure requires some non-collisional energy losses to fit the model to the data. This modeling needs improving to allow for pitch angle changes and for the variation of collision cross section as the target ionisation decreases, before firm conclusions are drawn.

Some events show more complex HXR structure though this is in part due to higher photon fluxes enabling detection of fainter components. An example is the extensively studied event of July 23, 2002, already shown in broad context in Fig. 1. Figure 3 shows that in the deka-keV HXR range the source comprises two bright footpoints, with hard spectra, and possibly a third, or at least one extended, footpoint, with a distinct fainter and softer source, possibly at or near the looptop. This event is one of those subjected most thoroughly so far to spectral image reconstruction [13] though that facet of RHESSI data reduction is still being refined.

Fletcher and Hudson [14] have studied the location and motion of RHESSI HXR 'footpoints' and compared them with those seen by TRACE in the XUV range and at other wavelengths (Fig. 4). This reveals a relatively complex situation, as yet to be properly understood. The HXR patches are rather large, being seen as extended even at this limited resolution ($\sim8''$) but nevertheless

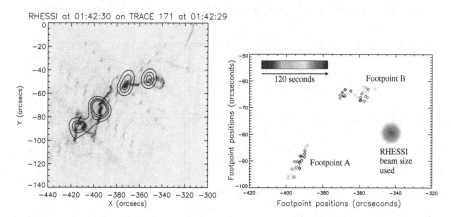

Fig. 4. Left: RHESSI 30–50 keV contours overlayed on a TRACE 171 Å image of the peak of the M5.4 flare of 14 March 2002 from [14]. The RHESSI images were reconstructed with CLEAN using grids 3 to 9 giving an angular resolution of $\sim8''$. At the available spatial resolution, there is a good correspondence between the HXR sources and the TRACE 171 Å kernels though the HXR patches are rather large, and at any moment occupy only a small part of the overall flare ribbon extent observed in TRACE. **Right**: Hard X-ray footpoint centroid motion in the flare from [14]. Footpoint locations derived from RHESSI 30–50 keV images are color-coded in time. Comparison with TRACE XUV images (cf. [14]; see also Figure on the left) reveal that the HXR source motions are perpendicular to as well as along the XUV flare ribbons indicating that the HXR footpoint progression is much more complex than expected from simple 2D reconnection models

at any moment occupied only a small part of the overall flare ribbon extent (Fig. 4, left), their centroids moving along it as the flare progressed (Fig. 4, right). The TRACE brightpoints are much smaller/better resolved and some of them more or less track the HXR patch centroids. Fletcher and Hudson interpret this motion as reflecting the progression of the momentarily reconnecting field lines which direct particles to the chromosphere. While this may be true, and their observed sizes the result of 'motion' of even smaller sources during the integration time, it is impossible in the conventional thick target model for HXR source electrons to be concentrated in regions anywhere near as small as a bundle of field lines near the very thin reconnection sheet. An intense burst requires a beam rate $n_b v_b A \approx 10^{36}\,\mathrm{s}^{-1}$ or a beam density $n_b \approx 10^{11}/A_{15}$ where $A = 10^{15} A_{15}\,\mathrm{cm}^2$ is the beam area which requires an impossible large n_b for $A \ll 10^{15}$ cm^2 ($\sim 1''$ square).

Veronig and Brown [15] discovered a new class of bright coronal HXR source in which the loop (top) emission is hard and dominates to high energies, with little or no emission from the footpoints, in sharp contrast to Masuda coronal sources where the footpoints still dominate – Fig. 5. Similar RHESSI events have been also studied by Sui et al. [16].

Veronig and Brown interpret this as due to a high loop density ($> 10^{11}$ cm^{-3}), consistent with the soft X-ray emission measure estimate $(EM/V)^{1/2}$, the coronal loop being then a collisionally thick target at electron energies up

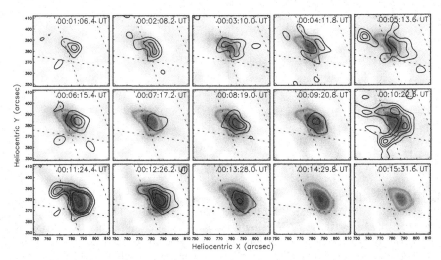

Fig. 5. RHESSI image sequence for the M3.2 flare of 2002 April 14/15, from [15]. The images show 6–12 keV, the contours 25–50 keV RHESSI maps reconstructed with CLEAN using grids 3 to 8 (except 7). The soft 6–12 as well as the hard 25–50 keV emission are concentrated near the loop top. Only during the impulsive rise (00:01–00:06 UT) and briefly during the late highest peak (00:10:22 UT), is weak footpoint emission detectable in the 25–50 keV band. Note that throughout the event the loop top is the predominant HXR source, whereas in 'normal' events footpoint emission prevails at high energies

to 50–60 keV. The high loop density is also consistent with conductive evaporation driven by collisional heating in the loop top. Such an interpretation had in fact previously been hinted at for Yohkoh data [17] but such events seem to be rare and more easily seen with RHESSI's large dynamic and spectral range.

An even more extreme class of high altitude source has been reported by Kane and Hurford [18] where there is an elevated, long lasting, HXR source seemingly 'detached' from chromospheric emission. Kane, McTiernan and Hurley [19] report a particularly interesting case of coronal HXR emission seen by RHESSI for an occulted flare behind the limb, but wholly seen by Ulysses which was behind the sun.

The coronal source yields a significant fraction of the total flare HXR flux showing the presence of copious fast electrons somehow confined high in the atmosphere, somewhat akin to the long duration high altitude γ-ray events studied earlier by Ramaty et al. in [20]. Much work remains to be done on the detailed quantitative modelling and physical interpretation of this class of HXR event, using the much more comprehensive data available from RHESSI, Nobeyama etc. than was possible in Kane's earlier ground-breaking stereo event studies [5].

3 Temporal Domain Discoveries and Issues

The most important temporal information in RHESSI data is bound to be in the evolution of the spatial and spectral characteristics, as opposed to global light curves in single energy bands. Unravelling the raw data to produce an X-ray 'multi-colour' movie at high spectral and spatial resolution is computationally very demanding and can only be even remotely contemplated for intense events with ample photons. However, a great deal can be gleaned from more rudimentary temporal information such as comparison of image sequences in two well separated energy bands, whole sun light curves as a function of energy over hitherto unexplored energy ranges, and comparison of light curves/image sequences at soft and hard X-ray energies with data at wholly different energies. In these categories, among the 'rethink' provoking RHESSI discoveries are the following.

There has long been interest in the possibility of 'nano'- or 'micro'- flares being an ongoing solar coronal phenomenon, possibly involved in coronal heating. Much research in this area has been statistical in character but some papers have addressed the physics of micro-events including their possible role in supplying mass to the corona, as well as heating it. Brown et al. [21] claimed that micro-events in loops were not hot enough to provide their emission measure increase by conductive evaporation of the chromosphere. They proposed that energetic electrons of around 10 keV might instead be responsible and predicted that RHESSI might detect frequent low energy 'hard' X-ray micro-events from the non-flaring sun. One of RHESSI's early discoveries was indeed that the 'non-flaring' sun exhibits micro HXR events of minutes' duration at intervals of several minutes – Fig. 6 [22]. The detailed physics of these remains to be investigated quantitatively but Krucker's movies (`http://sprg.ssl.berkeley.edu/~krucker`) show them to be related to XUV surges seen by TRACE, and to Type III radio bursts and associated plasma waves, originating near the boundaries of HXR sources.

The interplay of spectral, spatial and temporal information from RHESSI is particularly clearly shown in the study by Veronig et al. [23] of the Neupert effect. The empirical Neupert effect is that the SXR light curve of a flare is well correlated with the time integral of the HXR light curve [24, 25]. The canonical interpretation is that the SXR plasma is heated by cumulative energy input from HXR emitting fast electrons. The Neupert effect has been observationally established for the impulsive phase of many flares, but in general the correlations are far from being perfect [25, 26, 27, 28]. These deviations were interpreted as the effect of plasma cooling and/or grounded

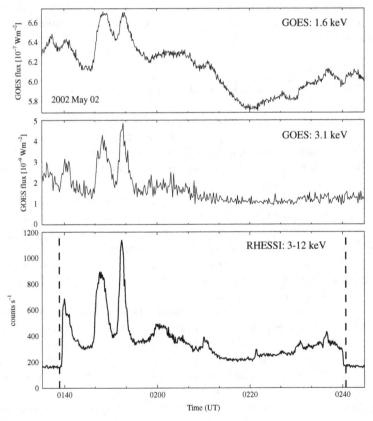

Fig. 6. Low energy HXR micro-events reported by [22] showing an early RHESSI discovery that even the non-flaring sun undergoes impulsive low energy but spectrally hard 'HXR' events quasi-continuously (intervals ∼ minutes)

on the idea that the *physical* Neupert effect should exist between the nonthermal and thermal energies and not between the HXR and SXR lightcurves. Veronig et al. quantitatively addressed this issue and suggested that the imperfect temporal correlation may be due to the fact that SXR flux depends on density and temperature, and not just on energy content, that HXR flux depends on beam spectrum as well as power, and that one must take account of plasma cooling by radiation and conduction during the event. There has of course been considerable theoretical/numerical work [29, 30] to see whether and how a Neupert effect is seen when one runs a model of the impulsive heating of a loop and follows its evolution allowing for hydrodynamics, evaporation, radiative and conductive cooling etc. The broad answer is, unsurprisingly, the Neupert effect being clearest when the loop takes longest to cool. However, such theoretical models are *not* the same thing as testing to see if the Neupert effect *in real data* can be physically attributed solely to beam heating of the hot gas after allowance for evaporation, cooling etc. That is, does the energetics of the beam input implied by real HXR data tally with the heating (or cooling) of the gas as inferred from real SXR data.

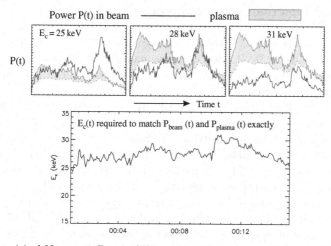

Fig. 7. Empirical Neupert effect in SXR and HXR light curves for the M3.2 flare of 2002 April 14/15 translated into SXR plasma power-in and power-requirement for a single loop, according to [23]. **Top panels**: Comparison of the actual power in the hot plasma P_{plasma} required to explain the observed SXR flux (minimum-maximum estimate: *shaded area*) and the electron beam power P_{beam} (*solid line*) calculated for different values of the low cutoff energy E_c. No value for E_c yields a good match between the $P_{plasma}(t)$ and $P_{beam}(t)$ curves. **Bottom panel**: If we allow the low cutoff energy E_c to change during a flare and see how it has to vary in order that $P_{plasma}(t)$ and $P_{beam}(t)$ derived from observations exactly match at each time step, then it is found that only small changes in E_c are necessary

Using a rather crude model of energy transport, but one which should correct the light curves in the right sense at least, Veronig et al. [23] studied this for several flares and found the surprising result that the power into the SXR plasma is less well correlated with the beam power than are the raw SXR and HXR light curves – Fig. 7. They then discuss reasons for this result including the possibility that models involving beam-heating of a single monolithic loop structure may be invalid. In line with the findings of Fletcher and Hudson [14] mentioned in Sect. 2, multiple small loop events might offer an alternative and more realistic description of the global flare emission. These, however, also face serious problems since the instantaneous electron rate (s^{-1}) is fixed by the HXR burst intensity and if one decreases the instantaneous area of injection to that of a small subloop elements one quickly reaches the point where there are simply not enough electrons for imaginable densities. In fact, filamentary beams cannot be less than around 0.1 of the loop radius thick for this reason, as well as such beams and/or their return currents being unstable. This is far thicker than the scale of current sheets.

4 Spectral Discoveries and Issues

Figure 8 shows dramatically the sea-change which RHESSI's Ge detectors have brought to flare HXR spectrometry, namely an increase in resolution from tens of keV to around 1 keV, enabling detailed spectral analysis of the bremsstrahlung continuum and resolution of individual γ-ray lines. The importance of this, emphasised for decades (e.g. [31]) lies in the fact that the mean source electron spectrum is essentially the derivative of the photon spectrum (deconvolved through the bremsstrahlung cross-section) and that the injected electron spectrum is the further deconvolution ($\sim$differentiation) of the mean source electron spectrum through particle transport smearing effects.

To see this we summarise the derivation of these relationships here.

In the approximation of isotropic emission, the hard X-ray photon spectrum $I(\epsilon)$ in solar flares is a convolution of the (density weighted volumetric) *mean source electron spectrum* $\overline{F}(E)$ and the cross-section $Q(\epsilon, E)$ for production of a photon of energy ϵ by an electron of energy E, viz.

$$I(\epsilon) = C \int_\epsilon^\infty \overline{F}(E)Q(\epsilon, E)\, dE ,$$ (1)

where C is a constant [3, 32].

Thus to find $\overline{F}(E)$ from $I(\epsilon)$ we have to solve/invert this integral equation which is always rather unstable to noise in $I(\epsilon)$. Put another way, the integral involved smears out features of $\overline{F}(E)$ in emitting the observable $I(\epsilon)$. A clear example is in the approximation (Kramers) $Q \sim 1/(E\epsilon)$ which leads to the explicit derivative solution

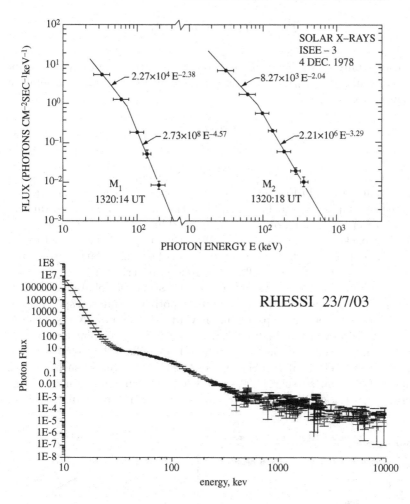

Fig. 8. Comparison of HXR spectral resolution pre- and post-RHESSI. The improvement between the 1980s (ISEE etc.) and now (RHESSI) in spectral energy points/bins from $\Delta\epsilon/\epsilon \sim 0.3$ to ~ 0.1–0.001 over the 10–1000 keV range is huge and has enabled the first systematic objective inference of electron spectra from their HXR bremsstrahlung spectra since it was first proposed by Brown [3]

$$\overline{F}(E) \sim -E \left[\frac{d(I\epsilon)}{d\epsilon} \right]_{\epsilon=E} \tag{2}$$

and differentiating data always magnifies high frequency noise [33].

In turn, $\overline{F}(E)$ is related to the electron injection rate spectrum $F_0(E_0)$ through the properties of electron propagation in the source. In the case of purely collisional transport the relation is [34]

$$\overline{F}(E) \sim E \int_E^{\infty} F_0(E_0)dE_0 \tag{3}$$

(so that for a pure power law $\overline{F}(E)$ is two powers harder than $F_0(E_0)$). Consequently details in $F_0(E_0)$ are further smeared out in $\overline{F}(E)$ and the solution for $F_0(E_0)$ for given $\overline{F}(E)$ is

$$F_0(E_0) \sim - \left[\frac{d(\overline{F}(E)/E)}{dE} \right]_{E=E_0} \tag{4}$$

so that $F_0(E_0)$ is sensitive to noise in $\overline{F}(E)$ and extremely sensitive to noise in $I(\epsilon)$ This is illustrated in Fig. 9 (left) where a 'top hat' feature in the injected electron spectrum is seen to be smeared in the mean source spectrum and smeared further in the final observed bremsstrahlung photon spectrum.

Working backward from the photon data thus requires the careful use of regularisation algorithms to suppress the effects of spurious data noise amplification [31] and a great deal of effort has gone into perfecting such methods [35, 36, 37, 38, 39]. Recent work has aimed at testing the reliability of these methods [40, 41] by applying them 'blind' to hypothetical photon spectra generated by models of electron spectra unknown to the data analysers. Overall all the methods prove to be highly reliable, as can be seen from the example in Fig. 9 (right) the only problematic regimes being where the source electron spectrum has low electron numbers in some energy range. In such regimes the photon spectrum at energy E is dominated by emission from electrons of energy $E \gg \epsilon$, rather than $E \simeq \epsilon$, and so is a rather poor diagnostic of

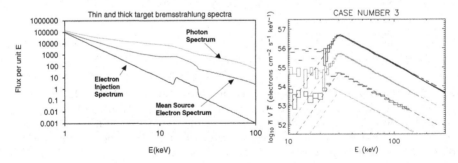

Fig. 9. Left: Two stage smearing of a 'top-hat' feature on a power-law injected electron spectrum $F_0(E_0)$ via collisional *thick target* transport to give mean source ('*thin target*') spectrum $\overline{F}(E)$ and via bremsstrahlung cross section to give the photon spectrum $I(\epsilon)$. **Right**: One early result of blind tests of spectral reconstruction algorithms for deriving mean source electron spectra from noisy photon spectra by regularisation/smoothing techniques. The four curves, displaced for clarity, are the results of four different algorithms [36, 38, 40, 42] while the *dashed curves* are the input function to be recovered. The target electron spectrum was unknown to the data 'inverters' till all results were in, so as to test the objectivity and consistency of the methods

electron fluxes at such energies. In practice photon spectra and hence electron spectra are, on the whole, quite steeply decreasing at all energies so that the electron spectra should be quite well recovered at most E for most events.

In a number of events, however, and most notably in the event of July 23, 2002, the observed HXR spectrum shows locally hard/small spectral index regions in the 30–60 keV range. When deconvolved, such photon spectra yield mean source electron spectra with a non-monotonic 'dips' of a kind hitherto totally unknown and of potentially great importance since their presence might rule out the canonical thick target model with purely collisional transport which can only produce a photon spectrum of local spectral index $\gamma \geq -1$ [34]. It is therefore of vital importance to test the reality of such features as originating in the flare electrons themselves rather than in some secondary process. To date, instrumental origins such as pulse pile up have not been entirely ruled out while a major issue is the contribution to the photon spectrum of photospheric back scatter [43] which is important in the 30–60 keV range. This arises from the Compton scattering and photoelectric absorption of downward directed photons from both free and bound electrons. Preliminary analysis of this effect by Kontar and others suggests that it can result in inference of a spurious electron spectral dip in the 30–40 keV range – Fig. 10 – but that the July 23, 2002 feature ∼50–60 keV might be too high in energy to be attributable to albedo.

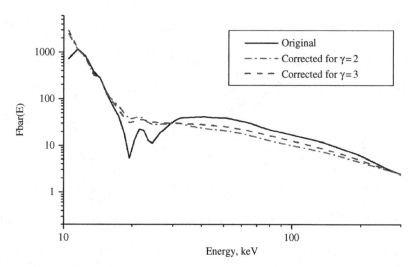

Fig. 10. Possible removal of inferred 'dip' in electron spectrum $\bar{F}(E)$ by albedo correction of photon spectrum [44]. When the observed photon spectrum is assumed to be solely due to primary bremsstrahlung emission the inferred electron spectrum $\overline{F}(E)$ for some RHESSI flares showed a dip around 30 keV. But when a correction was applied to remove the contribution from photospherically backscattered downgoing photons, the dip vanished

The importance of albedo in the observed signal also depends on the extent of the directivity (downward beaming) of the primary HXR source [45]. Even in the absence of albedo, source directivity affects the inferred slope and flux of the source electrons – Massone et al. [46] – Fig. 11 – thus much work remains to be done before we can be confident of inferred electron spectra.

A wholly different but equally intriguing RHESSI 'discovery', demanding rethinking of ideas, is the relationship of the mean spectral slope of HXR spectra from HXR source electrons at the sun to the slope of interplanetary electrons at the earth. RHESSI and WIND have allowed Krucker and Kontar (private communication) to confirm, with much improved data, earlier results of Lin et al. [47] and Lin [48, 49].

If all electrons are accelerated in the corona (as opposed to say separate acceleration sites in the corona and chromosphere) with spectral index δ then one would expect that: (a) upward interplanetary electrons would arrive at the earth with (scatter-free) index δ; (b) HXR's emitted from downward injection of such electrons into a dense collisional thick target [3] would produce HXRs of index $\delta - 1$; (c) HXR's emitted around the tenuous acceleration site would have thin target index $\delta + 1$. Krucker and Kontar, like Lin earlier, found that real flares are much closer to regime (c) than to regime (b).

This seeming violation of the predictions of the basic collisional thick target model is contrary to the RHESSI imaging results cited in Sect. 2 which

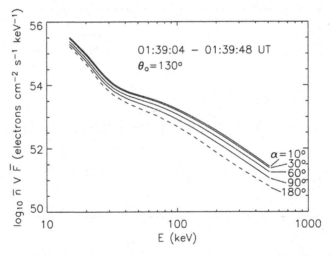

Fig. 11. Effect of bremsstrahlung cross-section anisotropy on the inferred $\bar{F}(E)$ for various degrees of source electron anisotropy, from [46]. What this shows is how the inferred source electron spectrum $\overline{F}(E)$ varies for a variety of assumptions on the anisotropy of the electron using the full anisotropic bremsstrahlung cross-section. Note that, as well as the shape of $\overline{F}(E)$ there is a large variation in the absolute value and hence in the total number and energy of the electrons. These results are for theoretical primary sources and are not confused by albedo effects

seems in broad support of that thick target model – i.e. we do see footpoints. It is possible that interplanetary electron propagation is not scatter-free or that thick target electrons undergo non-collisional losses but in either interpretation it is strange that the transport effects are just such as to yield data close to thin target situation (c). Time indeed for some rethinking!

5 Conclusions

RHESSI data constitute the greatest breakthrough in flare fast particle studies since the first HXR detectors were launched over 30 years ago. The results will pose 'rethink' challenges for an entire new generation of solar physicists, all the more so when considered in the wider context of multi-wavelength data, especially in the complementary radio regime to which CESRA is dedicated.

Acknowledgement

This work was supported by a PPARC Rolling Grant (JCB, EPK) and a Visitor Grant from the Royal Society of Edinburgh (AMV). We are grateful to Alec MacKinnon for help with the manuscript.

References

1. R.P. Lin, B.R. Dennis, G.J. Hurford, et al.: Solar Phys. **210**, 3–32 (2002)
2. B.R. Dennis, H.S. Hudson, S. Krucker: "Review of selected RHESSI solar results", in *The High Energy Solar Corona: Waves, Eruptions, Particles* (this volume)
3. J.C. Brown: Solar Phys. **18**, 489 (1971)
4. S. Masuda, T. Kosugi, H. Hara, S. Tsuneta, Y. Ogawara: Nature **371**, 495 (1994)
5. S.R. Kane: Solar Phys. **86**, 355–365 (1983)
6. R.P. Lin, R.A. Schwartz: Astrophys. J. **312**, 462–474 (1987)
7. G.J. Hurford, R.A. Schwartz, S. Krucker, R.P. Lin, D.M. Smith, N. Vilmer: Astrophys. J. Letts. **595**, L77–L80 (2003)
8. A.G. Emslie, J.A. Miller, J.C. Brown: Astrophys. J. Letts. **602**, L69–L72 (2004)
9. J.C. Brown, A.N. McClymont: Solar Phys. **49**, 329–342 (1976)
10. A.L. MacKinnon, J.C. Brown, J. Hayward: Solar Phys. **99**, 231–262 (1985)
11. J.C. Brown, M.J. Aschwanden, E.P. Kontar: Solar Phys. **210**, 373–381 (2002)
12. M.J. Aschwanden, J.C. Brown, E.P. Kontar: Solar Phys. **210**, 383–405 (2002)
13. A.G. Emslie, E.P. Kontar, S. Krucker, R.P. Lin: Astrophys. J. Letts. **595**, L107–L110 (2003)
14. L. Fletcher, H.S. Hudson: Solar Phys. **210**, 317–321 (2002)
15. A.M. Veronig, J.C. Brown: Astrophys. J. Letts. **603**, L117–L120 (2004)
16. L. Sui, G.D. Holman, B.R. Dennis: Astrophys. J. **612**, 546–556 (2004)
17. T. Kosugi, T. Sakao, S. Masuda, H. Hara, T. Shimizu, H.S. Hudson: "Hard and Soft X-ray Observations of a Super-Hot Thermal Flare of 6 February, 1992", in *Proceedings of Kofu Symposium*, eds. S. Enome & T. Hirayama (NRO Report 360, 1994), pp. 127–129

18. S.R. Kane, G.J. Hurford: Advances in Space Research **32**, 2489–2493 (2003)
19. S.R. Kane, J.M. McTiernan, K. Hurley: American Astronomical Society Meeting Abstracts **204** (2004)
20. R. Ramaty, N. Mandzhavidze, C. Barat, G. Trottet: Astrophys. J. **479**, 458 (1997)
21. J.C. Brown, S. Krucker, M. Güdel, A.O. Benz: Astron. Astrophys. **359**, 1185–1194 (2000)
22. S. Krucker, S. Christe, R.P. Lin, G.J. Hurford, R.A. Schwartz: Solar Phys. **210**, 445–456 (2002)
23. A.M. Veronig, J.C. Brown, B.R. Dennis, R.A. Schwartz, L. Sui, A.K. Tolbert: Astrophys. J. **621**, 482–497 (2005)
24. W.M. Neupert: Astrophys. J. Letts. **153**, L59–64 (1968)
25. B.R. Dennis, D.M. Zarro: Solar Phys. **146**, 177–190 (1993)
26. T.T. Lee, V. Petrosian, J.M. McTiernan: Astrophys. J. **448**, 915–924 (1995)
27. J.M. McTiernan, G.H. Fisher, P. Li: Astrophys. J. **514**, 472–483 (1999)
28. A. Veronig, B. Vršnak, B.R. Dennis, M. Temmer, A. Hanslmeier, J. Magdalenić: Astron. Astrophys. **392**, 699–712 (2002)
29. P. Li, A.G. Emslie, J.T. Mariska: Astrophys. J. **417**, 313–319 (1993)
30. P. Li, J.M. McTiernan, A.G. Emslie: Astrophys. J. **491**, 395 (1997)
31. I.J.D. Craig, J.C. Brown: *Inverse Problems in Astronomy* (Adam Hilger, 1986)
32. J.C. Brown, A.G. Emslie, E.P. Kontar: Astrophys. J. Letts. **595**, L115–L117 (2003)
33. E.P. Kontar, A.L. MacKinnon: Solar Phys. **227**, 299–310 (2005)
34. J.C. Brown, A.G. Emslie: Astrophys. J. **331**, 554–564 (1988)
35. A.M. Thompson, J.C. Brown, I.J.D. Craig, C. Fulber: Astron. Astrophys. **265**, 278–288 (1992)
36. C.M. Johns, R.P. Lin: Solar Phys. **137**, 121–140 (1992)
37. M. Piana: Astron. Astrophys. **288**, 949–959 (1994)
38. M. Piana, A.M. Massone, E.P. Kontar, A.G. Emslie, J.C. Brown, R.A. Schwartz: Astrophys. J. Letts. **595**, L127–L130 (2003)
39. A.M. Massone, M. Piana, A. Conway, B. Eves: Astron. Astrophys. **405**, 325–330 (2003)
40. E.P. Kontar, M. Piana, A.M. Massone, A.G. Emslie, J.C. Brown: Solar Phys. **225**, 293–309 (2004)
41. E.P. Kontar, A.G. Emslie, M. Piana, A.M. Massone, J.C. Brown: Solar Phys. **226**, 317–325 (2005)
42. G.D. Holman: Astrophys. J. **586**, 606–616 (2003)
43. R.C. Alexander, J.C. Brown: Solar Phys. **210**, 407–418 (2002)
44. E.P. Kontar, A. MacKinnon, R.A. Schwartz, J.C. Brown: Astron. Astrophys. **446**, 1157–1163 (2006)
45. S.R. Kane, K.A. Anderson, W.D. Evans, R.W. Klebesadel, J.G. Laros: Astrophys. J. **239**, L85–L88 (1980)
46. A.M. Massone, A.G. Emslie, E.P. Kontar, M. Piana, M. Prato, J.C. Brown: Astrophys. J. **613**, 1233–1240 (2004)
47. R.P. Lin, R.A. Mewaldt, M.A.I. van Hollebeke: Astrophys. J. **253**, 949–962 (1982)
48. R.P. Lin: Solar Phys. **100**, 537–561 (1985)
49. R.P. Lin: Advances in Space Research **13**(9), 265–273 (1993)

Small Scale Energy Release
and the Acceleration and Transport
of Energetic Particles

Hugh Hudson[1] and Nicole Vilmer[2]

[1] SSL/UCB Berkeley, CA USA 94720
   hhudson@hhudson@ssl.berkeley.edu
[2] LESIA, Observatoire de Paris, F-92195 Meudon, France
   nicole.vilmer@obspm.fr

Abstract. We report on results presented at the sessions of Working Group 1 at CESRA 2004, which covered the topic area of the title of this paper. The working-group participants are listed in the Appendix, and the topics discussed have been brought together in several general areas of focus. The emphasis on the discussion is from the point of view of radiophysics. We organize the material by presenting new constraints imposed by the recent high-energy and radio observations. We note though that multi-wavelength knowledge is generally vital in understanding all of the phenomena involved. The new constraints include exciting new millimeter-wave discoveries, among others. We then place these observations into the framework of our knowledge of the acceleration and propagation of high-energy particles, and of their radio emission mechanisms. The RHESSI[1] results are the most distinctive in this time frame, and they have made possible several new advances.

1 Introduction

The techniques of solar radio astronomy have historically led the way in our studies of the non-thermal behavior of the solar corona. In recent decades X-ray and EUV observations have begun to approach the resolution (the arc sec range) of the radio observations at mm/cm wavelengths (e.g. those of the VLA[2] and NoRH[3]) and we now find ourselves in a happy era in which these very different wave bands can all contribute to our understanding.

This article discusses material presented in the CESRA 2004 working-group sessions on the subject named in the title of this chapter; please see the CESRA 2001 proceedings [1] for continuity. Although we deal with multi-wavelength views of these topics, our perspective is that of solar radiophysics

---

[1] Reuven Ramaty High-Energy Solar Spectroscopic Imager (space observatory).
[2] Very Large Array (Socorro, New Mexico).
[3] Nobeyama Radio Heliograph (Nobeyama, Japan).

H. Hudson and N. Vilmer: *Small Scale Energy Release and the Acceleration and Transport of Energetic Particles*, Lect. Notes Phys. **725**, 81–103 (2007)
DOI 10.1007/978-3-540-71570-2_5   © Springer-Verlag Berlin Heidelberg 2007

(see also [2] for recent reviews and discussions in the context of the FASR[4] program). We have organized some of the new material from CESRA 2004 in these areas into sections on new observational constraints (Sect. 2, dealing with new X-ray and radio inputs), on particle acceleration and propagation (Sect. 3), and on radio emission mechanisms (Sect. 4).

The essence of the radio observational technique lies in its ability to sense the presence of energetic electrons remotely by a variety of emission mechanisms.

Particle acceleration frequently accompanies the coronal restructuring involved in a flare or a coronal mass ejection (CME), and general consensus holds that multiple kinds of particle acceleration are at work. Radio astronomers have also observed for years the production of suprathermal electrons in association with active regions even in the absence of flares (e.g., metric type I noise storms). With radio techniques we can perform imaging spectroscopy over a many-decade span of the frequency spectrum, observing free-free continuum, plasma emissions or synchrotron radiation. The exciting thing now is to place these remote particle observations into the context of the plasma motions and heating observed at other wavelengths, and to relate them quantitatively to the coronal magnetic field. The association with solar energetic particles (SEPs) as observed in situ also has become a more important undertaking as the imaging and magnetographic data improve in quality. Radio observations in the future will indeed contribute substantially to our understanding of the solar corona.

Only a handful of radio observatories currently have imaging multifrequency observational capability and are devoted to continuous solar observation. Others have occasional programs of solar observation. The VLA and the GMRT[5] are examples of the latter, and we hope that ALMA[6] will be another in the future. Table 1 summarizes some of the currently-available observational capabilities in the radio domain. The radio spectrum is vast and the available facilities highly specialized in many cases. We apologize for the sketchiness of Table 1, which really ought to be replaced by a four-dimensional graphic (spatial resolution, temporal resolution, spectral resolution, polarization capability).

Future major new solar-dedicated observing facilities (specifically FASR and the Chinese Radioheliograph) will greatly extend the table of capabilities. Furthermore the importance of flare observations in the mm-submm range, as described below in Sect. 2.2, strongly motivates observations with better spatial resolution, possibly with ALMA. The existing SST[7] will be upgraded to perform observations in the sub-millimeter domain at 850 GHz and in the near infrared (7-14μ, or 21.5-43 THz). A new concept for space observations

---

[4] Frequency-Agile Solar Radiotelescope (in development).
[5] Giant Metrewave Radio Telescope (Pune, India).
[6] Atacama Large Millimeter Array (in development).
[7] Solar Submillimeter Telescope (El Leoncito, Argentina).

Table 1. Solar radio astronomy: current capabilities of solar-dedicated instruments

Range	Image resolution	Time resolution	Spectral resolution
submm	Few′ (SST)[a]	1 ms	Fixed frequencies 212 GHz-405 GHz
mm	Few″ (NoRH)	1 s	Fixed frequencies 17 GHz-34 GHz
cm	10″ (OVRO[b])	50 ms	Few % 1-18 GHz
cm	15″ (SSRT[c])	>14 ms	Fixed frequency 5.7 GHz
dm/m	~1′ (NRH[d])	125 ms	5-10 frequencies 150-450 MHz

[a] Centroiding, rather than true imaging
[b] Owens Valley Radio Observatory [3]
[c] Siberian Solar Radio Telescope [4]
[d] Nançay Radio Heliograph [5]

has also been proposed to carry far-infrared observations (35 and 150μ, or 2 and 8.6 THz) together with γ-ray observations (MIRAGES; [6]). These new capabilities will greatly strengthen the quantitative interpretation of many observational properties currently known only morphologically and will surely lead to great progress.

Finally, we note the crucially important space-borne observations at EUV, X-ray, and γ-ray wavelengths. SOHO[8], TRACE[9], and RHESSI figure prominently in the list of currently-operating spacecraft with broad capabilities. RHESSI provides high-resolution γ-ray spectroscopy, as well as imaging, and higher-resolution spectroscopic hard X-ray observations are also now becoming available from SMART-1 [7] and GSAT-2 [8] at lower energies. Imaging spectroscopy at high resolution is also possible in principle with microcalorimeter arrays, which have already been deployed in space for non-solar observations (e.g., [9]). Solar X-ray astronomers should take note and apply these techniques, with high-resolution imaging, to solar observations as well.

2 Some New Observational Constraints

2.1 Hard X-rays

Microflares

Flare occurrence generally follows a power law in total energy, as approximately shown in a wide variety of observations (e.g., [10]). The nomenclature is confusing; "microflare" refers to a tiny but otherwise undistinguished solar flare with total energy on the order of 10^{26} ergs – one millionth of a major flare at 10^{32} ergs. "Nanoflare" on the other hand refers to a different physical

---

[8] Solar Heliospherical Observatory (space observatory).

[9] Transition Region And Coronal Explorer (space observatory).

process, hypothesized by Parker [11] to explain coronal heating in terms of ubiquitous tiny non-thermal energy releases. The nanoflares, on this interpretation, would have an occurrence distribution function so steep that individual events would not be individually recognizable. Thus one could not observationally distinguish flares/microflares from nanoflares, except statistically or indirectly from their consequences. From other perspectives there appears to be a continuous spectrum of flares of all magnitudes. Figure 1 (left) shows that RHESSI microflare locations strongly tend to occur in active regions. Figure 1 (right) shows RHESSI thermal parameters for a similar sample of events, revealing higher temperatures (or smaller emission measures) than obtained from X-ray emission-line spectroscopy [12].

For the first time for such small events, RHESSI can trace the hard X-ray spectrum to photon energies of a few keV, well below the commonly-assumed low-energy cutoff at 20–25 keV. Because the spectrum is a soft power law, this means larger total energies than might be expected for these weaker events. The non-thermal energies of these tiniest events are surprisingly large even given a possible RHESSI bias towards higher temperatures (Fig. 1, right [13]), which could imply that the RHESSI thermal source contains only a fraction of the emission measure for the smaller events.

Ribbon Behavior

The Hα flare ribbons (which contain the hard X-ray footpoints) mark the photospheric/chromospheric boundary of the flare's magnetic flux tubes (see the cartoon in Fig. 2 [14]). This type of sketch adequately reveals the connectivity of the flare loops but does not describe the open fields linking the flare to the large-scale corona and solar wind, upon which SEPs must travel. Many

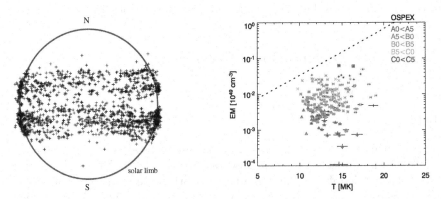

Fig. 1. Left: A map of RHESSI microflare positions, taken from an early 3-month sample. **Right**: Emission measure vs temperature for a smaller sample [12]; the *dotted line* shows the general correlation for flares [13]

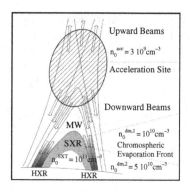
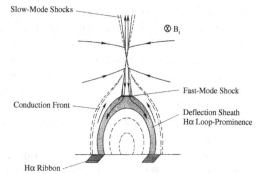

Fig. 2. Cartoon scenarios for magnetic reconnection in solar flares (*left*, from [14]; *right*, from [20]). The two views show essentially the same geometry, but the right-hand shows various shock waves that may form and be important for particle acceleration

variants of this "CSHKP" cartoon[10] have been published, and we generically call them "the reconnection model" here. This model is essentially motivated by the observations and provides a framework for discussion of eruptive or quasi-eruptive flares (e.g., [15]), but the theories remain more descriptive than predictive at present.

In the thick-target model for flare energetics the 10-100 keV electrons dominate the flare energy release, and we can trace their presence not only with hard X-rays but also (at higher resolution) with the UV and EUV imaging by TRACE [16, 17]. These observations show that the footpoints consist of bright kernels of emission with small spatial scales. This illustrates the complexity of the different scales involved in a flare/CME event; recent observations have tied the impulsive phase of the solar flare with the acceleration phase of a CME [18, 19].

Radio observations have not given us much information about the flare footpoint regions themselves, owing to absorption (free-free or gyrosynchrotron self-absorption) in the overlying flare and surrounding active-region atmosphere. At submillimeter wavelengths we do not have so much absorption but angular resolution thus far is relatively poor; we discuss new discoveries here in Sect. 2.2 below.

The key observational evidence for a model such as that of Fig. 2 lies in the behavior of the ribbons (seen in Hα and many other wave bands) and hard X-ray footpoints of the closed coronal magnetic loops. The footpoints indeed reflect the ribbon structure in that the hard X-ray sources are embedded in the ribbon regions [21, 22, 23]. However it is puzzling that these hard X-ray footpoints often only appear as a pair of compact sources with much smaller extent than the Hα ribbons themselves. The outer edges of the expanding flare ribbons should have broad line profiles as a result of the energy release

---

[10] Carmichael, Sturrock, Hirayama, and Kopp-Pneuman.

envisioned in the standard model (e.g., [24]); these footpoint regions should show the "explosive" evaporation driven by non-thermal electrons closely associated with the dynamics of the reconnection. Given the time variability observed in X-rays, new estimations of Hα line profiles and energy deposition as a function of depth resulting from time-modulated high-energy power law electron beams are being developed (e.g., Varady et al., this workshop).

The impulsive-phase emissions of a solar flare appear throughout the spectrum, via different emission mechanisms often related to non-thermal particles. These emissions typically correlate well temporally (e.g., [16] for the example of the Bastille Day flare of 14 July 2000), but their spatial behavior differs. Figure 3 (left) shows how the UV footpoint sources behaved in an M8.5 flare observed by TRACE on 17 July 2002; this reveals many simultaneous sources that move within the ribbons. In contrast to this, the typical pattern of hard X-ray emission consists of a single dominant footpoint in each ribbon, which often moves along the ribbon as the flare progresses [22, 25]. This difference in image morphology has not been explained and seems inconsistent with the usual thick-target model by which we identify the Hα ribbon and hard X-ray footpoint sources as the result of energy deposition by particles accelerated in the corona.

The apparent motions of hard X-ray footpoints sometimes appear to be consistent with the expectation from the cartoons (e.g., Fig. 2) in that they separate with time. Such a relationship has been sought in several studies, with varying degrees of success. In fact the majority of events display footpoint motions parallel to the ribbon elongation [16, 23, 25, 26, 27], rather than perpendicular to it. RHESSI observations have however shown a convincing correlation of properties expected in the magnetic reconnection model, illustrated in Fig. 3 (right), for the X-class flare of 23 July 2002 (see also [23]). In the model, the coronal magnetic field not only contains the energy to be

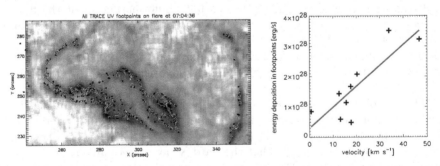

Fig. 3. Left: TRACE image of an M8.5 flare on 17 July 2002, showing the locations of UV bright points (footpoint sources) embedded in the ribbon structures. Several UV footpoints can exist simultaneously, and they move regularly through the ribbon envelope with a tendency to reflect magnetic features [17]. **Right**: Correlation between apparent footpoint velocity and hard X-ray flux for the flare of 23 July 2002

released (in the form of excess $B^2/8\pi$), but also guides the energy (in the form of fast particles) into the footpoints of the coronal magnetic loops that contain the released energy. However the majority of events show footpoint motions parallel [28] to the ribbons and in various senses, behavior not explainable in a 2D cartoon representation.

Evidence of Current Sheets

A series of RHESSI flares observed in April 2002 suggests some further confirmation of the reconnection model, while at the same time revealing unexpected behavior. We illustrate the first of these two points in Fig. 4, based upon the events described by Sui et al. [29, 30]. The left panel shows contours from three narrow bands in the range 6–20 keV, with double coronal sources apparently stretched out in the radial direction. The higher-energy (hence hotter) sources are the innermost pairs of contours. The authors interpret this phenomenon as a hot and dense current sheet forming in the corona above the flare loops, following the idea originally proposed by Syrovatskii (e.g., [31]). Such a current sheet would be a reasonable expectation for the reconnection model (Fig. 2), except that present-day theory cannot predict the temperature or density of the reconnecting structure. The right panel of Fig. 4 shows height-vs-time plots of loop-top and coronal sources. The coronal source – the upper anchor of the current sheet – initially remains stationary, even during the intense energy release of the impulsive phase of the flare, while the loop-top source unexpectedly moves downward (see [32] for an additional example of this behavior). These observations seem to contradict the

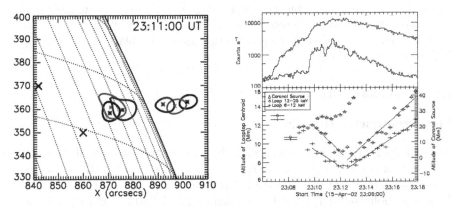

Fig. 4. RHESSI observations of apparently thermal hard X-ray sources in a flare of 15 April 2002. The panel at the *left* shows the footpoints of the main flare loop as +'s, while the contours show X-ray emission in three narrow bands in the range 6-20 keV (see [29]). The panel on the *right* shows the source motions, which reveal an unexpected *downwards* motion of the loop-top source during the initial phases of the flare

reconnection model, which associates energy release directly with plasmoid eruption (e.g., [33]), but they could be generally consistent with the need to extract energy from the coronal magnetic field [34].

The newer observations of coronal sources can be compared with the well-known *Yohkoh* observation of the "Masuda flare" [35], in which a hard X-ray source appeared above the soft X-ray loop top during the impulsive phase of a flare on 13 January 1992. This single observation gave a great deal of momentum towards the acceptance of the reconnection model, but the new RHESSI data on the behavior of coronal sources do not generally follow this pattern. Instead of a single non-thermal "Masuda" source, one sees paired coronal thermal sources in the events of Fig. 4. Coronal non-thermal sources do occur in the RHESSI data but in different configurations (e.g., [19, 36, 37]) that have not yet been systematized.

Thick-Target Coronal Sources

The same (Fig. 4) series of flares displayed an interesting hard X-ray behavior, as described by Brown et al. (this volume): the RHESSI data imply that the corona itself can be dense enough to stop non-thermal electrons, thus leading to a "coronal thick target hard X-ray non-thermal source" observed up to 25 keV with little or no emission from the footpoints ([37]; see also [38, 39]). This behavior would result from emission by non-thermal electrons in a high-density loop consistent with the one deduced from the soft X-ray emission measure. The usual assumption, based on standard semi-empirical models of the solar atmosphere, would be that the electrons should penetrate to the chromosphere before losing their energy to collisions. This assumption may still be true in most cases, but the interpretation of these flares suggests that the loops may have become too dense for this to happen. We can speculate that in such cases the electron acceleration may take place in a relatively high-density coronal region.

2.2 Radio Observations

Meter-centimeter Domain

The extended flare of 2003 Nov. 3 provides an excellent example of the manner in which the new data can describe a complicated flare/CME event (see the right panel of Fig. 5). The study presented by Dauphin et al. [40] uses Nançay radioheliograph, LASCO coronagraph, and RHESSI data (cf. Maia, this Workshop; Vršnak, this Workshop). The development of this flare/CME puts it in the category of "extended events", in which major coronal disturbances appear some time after the initial impulsive development of the flare. The time variation of this flare observed above 100 keV by RHESSI presents two broad series of bursts separated by a period of 4 minutes in which only

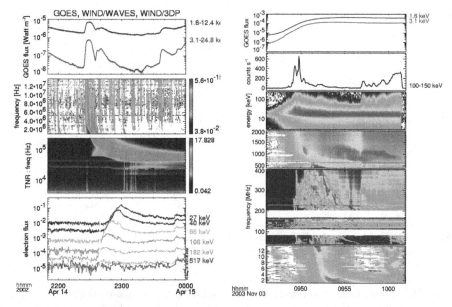

Fig. 5. *Left* Type III burst and interplanetary electron event; *right* from top to bottom: Time evolution of the GOES X-ray flux and of the X-ray RHESSI counts in the 100–150 keV energy band, RHESSI X-ray spectrogram, radio composite spectrum observed between 2 GHz to 1 MHz by PHENIX-2 (ETH Zürich), OSRA (AIP Potsdam) and the WIND/WAVES experiment. Note the continuum enhancement at 09:57 UT corresponding to the second phase of energy release observed in hard X-ray wavelength range [40]; (Krucker, this Workshop)

X-ray emission below 100 keV is observed. The first part of the X-ray emission is as usual associated in the dm/m domain with type-III-like bursts. The second part (after 09:57 UT) is mainly associated with a strong continuum emission in the whole range from 2 GHz to almost 200 MHz. A decimetric/metric type II emission is observed between the two parts of the hard X-ray emission starting at an unusually high frequency of 600 MHz. A fast (1420 km/s) coronal mass ejection is observed by the LASCO coronagraph on SOHO.

In this event, the "extended" phase begins extremely suddenly, at approximately 09:57 UT (Fig. 5, right); the onset crosses many wavebands (from 200 MHz through 89 GHz [40] and into the hard X-rays). The observations suggest that gyrosynchrotron emission is the prevailing emission mechanism even at decimetric wavelengths for the broad-band radio emission. The simultaneity of this broad-band is much sharper than typical CME time scales, limited by the local Alfvén speed or some low multiple of it, and strongly suggests non-thermal particles as the coronal energy transport in this case. RHESSI can image the HXR sources in both phases, as shown in Fig. 6. They are dominated by footpoints above 50 keV just as in the impulsive phase (see

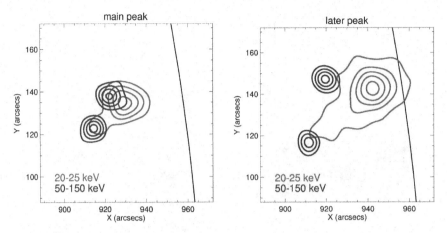

Fig. 6. RHESSI images of the early and late phases of the 2003 Nov. 3 event (Krucker, this Workshop; see also [40])

also [41]); these data establish that "extended flares" also have dominant footpoint emission in hard X-rays despite their strong coronal effects. However, the footpoint separation is larger in the late phase, as is the extent of the 20–25 keV source. The link with the CME onset should be further examined.

A New Observing Window for Flares: mm to submm

Submillimeter observations of flares (presented by Lüthi and by Trottet) provide new diagnostics for analyzing high energy electrons in solar flares. Indeed if emitted by gyrosynchrotron emission they require ultrarelativistic electrons (> few MeV) to explain them. Thus such data point to the most extreme particle acceleration processes. Prior to this millennium, however, no *submillimeter* and few mm-wave observations of flares had been reported (e.g., [42]). We now have independent submillimeter observations of flares from two observatories: the SST at El Leoncito in Argentina [43] and the Köln Observatory for Submillimeter and Millimeter Astronomy (KOSMA) at Gornergrat in Switzerland [44]. The instruments use substantially different techniques, combining those of radio and infrared astronomy. The higher frequency at El Leoncito, 405 GHz, corresponds to a wavelength of 740 μm, and an additional atmospheric window occurs at about half that wavelength. The SST is being upgraded to perform there (850 GHz) and in the far infrared (7–14 μm or 43–21.5 THz) (Trottet et al., this Workshop). The increasing opacity of the terrestrial atmosphere as one goes to short wavelengths, due largely to water vapor, makes it essential to observe from a high, dry site with adequate spatial reference to cancel out the atmospheric fluctuations (e.g., [45]). The large quiet-Sun brightness at high frequencies also implies the use of an interferometric or spatially chopping scheme for background cancellation.

The opening of the new spectral windows above 200 GHz for solar flares has provided some unexpected results. The spectral upturn seen at the highest frequencies (212 and 405 GHz at El Leoncito [43]; 230 and 345 GHz at Gornergrat [44]), as seen in the flux-density spectrum (Fig. 7), is not consistent with an optically thin thermal source, nor with the high-frequency extension of the optically-thin gyrosynchrotron emission of energetic electrons observed below 100 GHz. Likewise the increase is inconsistent with synchrotron self-absorption if non-thermal. For thermal emissions, we are thus likely to be viewing thermal sources not physically located in the solar corona, but rather in denser atmospheric layers (see [47]). If this is the case, then substantial new theoretical work will be required. We speculate that the RHESSI results on the 0.511 MeV γ-ray line width [48] (see also Dennis et al., this volume) also require a new treatment of the lower solar atmosphere during flare conditions, especially in view of the recent discovery of near-IR emission from flares [49]. In the case of nonthermal emissions, the > 200 GHz emission may arise from optically-thick synchrotron emission from relativistic electrons in a source different from the one emitting at low frequencies, free-free emission from the chromosphere due to energy deposited by electrons or protons or by synchrotron emission from pion-decay positrons. This last process, first described by Lingenfelter and Ramaty [50] could be reconsidered for the high frequency observations given the possible observation of π_0-decay γ-rays from flares showing a spectral increase above 200 GHz.

Imaging in the submillimeter domain remains limited, by diffraction, to the arc-minute resolutions. However both the El Leoncito and Gornergrat observations involve multiple feeds, providing for crude image centroiding and

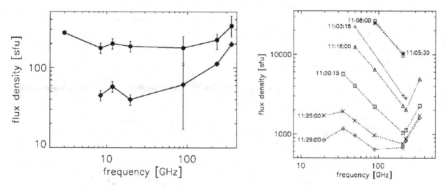

Fig. 7. Microwave/submm spectra from (*left*) an X2.1 flare on 12 April 2001, observed with the KOSMA telescope on Gornergrat [44], and (*right*) an X1.7 flare on 28 October 2003 [46]. The spectra show a surprising increase at the highest frequencies (230 and 345 GHz), inconsistent with a thermal source or with the extension towards high frequencies of the optically thin part of the gyrosynchrotron emission observed below 100 GHz

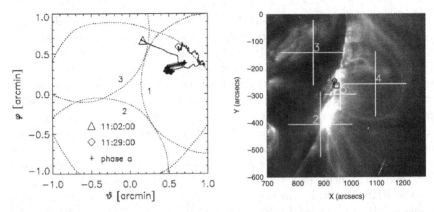

Fig. 8. Left: source positioning for the flare of 28 October 2003 during different periods of the flare [46], showing the layout of the multiple feeds of the KOSMA telescope. **Right**: source positioning by SST for the flare of 4 November 2004 [43]; the large crosses show the mapping of the feed locations on a TRACE image, and the symbols show centroid locations for different phases of the flare

size determination [46, 51]. The flare centroids are determinable with arc-second resolution and may show systematic apparent motions, as seen in the examples of Fig. 8.

As the right side of Fig. 8 suggests, the centroid locations for a major limb flare occurred at low altitudes. This image localization reinforces the idea derived from the spectral turn-up that the THz emission is concentrated in the low atmosphere. Note however that this flare was characterized by dense coronal loops that were bright enough to appear in white light even projected

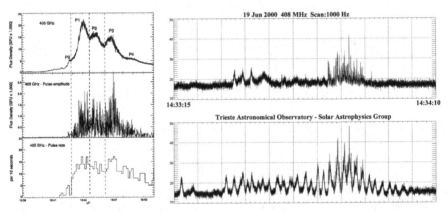

Fig. 9. Left: Rapid variability detected at the highest frequency observed at El Leoncito for the flare of 4 November 2004 [43]. The panels, from *top to bottom*, show the flux, the pulse amplitude, and the pulse rate. **Right**: Example of extremely fast variations at 408 MHz in observations from the Trieste Solar Radio System (Magdalenič, this Workshop)

against the dark sky [52], and if the submm sources included a coronal contribution the centroid locations might need re-interpretation. Higher angular resolution at mm and submm wavelengths, with true imaging, would therefore help to clarify the nature of the new spectral component.

Lastly the time variability of the new mm-wave sources also may offer some new surprises. The El Leoncito observations appear to show rapid (100–500 msec) modulations [43]. These modulations could in principle reflect an intermittent energy release best visible at the highest particle energies (hence the shortest wavelengths in the synchrotron spectrum). Figure 9 (left) shows an example of this kind of variability.

Ejecta and Fine Structures

The event of 2003 November 3 (Sect. 2.2) was also an eruptive event, and the ejecta could be followed by several instruments in the low corona [53]. We still

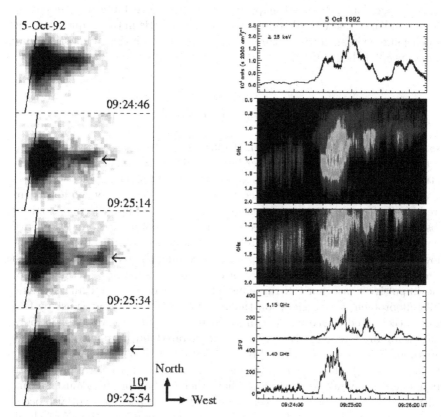

Fig. 10. Drifting pulsating structures (DPS) of 5 October 1992 observed in *Yohkoh* SXT soft X-ray images (*left*) and with the Ondřejov spectrograph (*right*). The upper panel shows BATSE hard X-rays >25 keV

have no theoretical consensus on the reason for the coronal loss of equilibrium that produces a coronal mass ejection, but current data have shown it to be closely related in time to the impulsive phase of the associated flare [18] when observable. The RHESSI events of 2002 April 21 and 2002 July 23 provide well-observed RHESSI examples of ejective flares [19, 36].

Now decimetric spectrographs show a class of emissions known as "drifting pulsating structures (DPS)," which can be imaged in hard X-rays (e.g., [54]). These sources reflect plasmoid ejection and may be associated at metric wavelengths with type II radio bursts. Karlický (this Workshop) proposes that such structures map the magnetic field reconnection responsible for the flare energy release, and furthermore that the decimetric time variability (the pulsations) correspond with reasonable time scales for bursty reconnection. Figure 10 shows an example, with good time coincidence between DPS occurrence, plasmoid ejection, and hard X-ray emissions. Super-fast structures have also been reported with characteristic time scales of a few tens of ms and narrow bandwidth around 10 MHz in the dm wavelength range (Fig. 9, right) (Magdalenič, this workshop). They probably reveal intermittent energy release and are preferentially observed around 600 MHz in the frequency range which could be imaged with future instruments such as FASR or a forthcoming Chinese radioheliograph.

3 Acceleration and Propagation Mechanisms

As stated by Burgess (this volume), "Collisionless shocks are a key component in astrophysical systems to transfer bulk flow kinetic energy to a small population of highly energetic particles and it is a truth universally acknowledged that shocks are effective particle accelerators." Shocks in many configurations are indeed often cited to explain various populations of non-thermal particles observed in the interplanetary medium and in the corona in connection with flares and CMEs. The universal role of electron acceleration by shocks has however been questioned (e.g., [55, 58]) concerning the real efficiency of such a mechanism to produce flare energetic electrons interacting at the Sun for the impulsive-phase hard X-ray emission. Flares pose indeed an especially difficult problem for acceleration theory, since so much energy must be tied up in accelerated electrons (as confirmed by RHESSI observations [36, 56]), and even more interestingly, also in the accelerated ions responsible for γ-ray emissions [57] (as also confirmed by RHESSI observations of e.g. the 23 July 2002 event [36]). It could be difficult to invoke shock acceleration with so high an efficiency in this context, so that other mechanisms also must be considered. Relativistic particle acceleration of escaping particles can also occur in the corona far from the flare site, not in connection with the CME shock and not coincident in time with the impulsive phase. How would a large-scale shock be created and efficiently accelerate particles in such a circumstance?

The traditional "thick target" model envisions a "black box" acceleration site in the low corona, related in one way or another to magnetic reconnection, from which beams of electrons stream up and down and produce radio and HXR emissions preferentially. These electrons have a power-law spectral distribution that may extend into the γ-ray energy range. There is substantial uncertainty about the location of the "black box" and also about whether the acceleration properties can be easily disentangled from particle propagation effects (viz., the currently developing "collapsing trap" ideas [31, 59, 60, 61]). This is one of the reasons why studying flare ribbon development, searching for evidence of current sheets, and observing thick-target coronal sources (Sect. 2.1) is so crucial. Brown et al. (this volume) provide a modern view of this model and note directions in which it may be changing based on new features observed by RHESSI. In principle an inverse-theoretical calculation can directly determine the nature of the source electron energy distribution from the X-ray spectra [62, 63], aided greatly by the improved energy resolution RHESSI provides.

The relationship between magnetic reconnection and particle acceleration in solar flares is presently not understood. Many authors discuss direct acceleration in the current sheet (represented by the X point in Fig. 2; see e.g., [64, 65, 66, 67]). Electron and proton spectra are computed and it is shown that increasing the longitudinal ("guide") magnetic field in the RCS increases the acceleration efficiency (e.g., [64]). Another interesting prediction of test-particle simulations is to show that the inclusion of a finite guide field introduces some asymmetry in the particle propagation, implying that electron numbers and spectra could be different from one footpoint to the other [65]. Changing the sign of the particle's charge should reverse the predominant footpoint; protons and electrons could be accelerated to different footpoints as also shown by Zharkova and Gordovskyy [68]. It should be further investigated whether this effect may explain the different electron and ion interactions sites observed with RHESSI [69] (see also Brown et al. and Dennis et al., this volume).

Other possibilities less directly related to the reconnection process also exist, specifically in the large-scale shock waves associated with the plasma flows (fast-mode or slow-mode, or in turbulence excited by these flows (e.g., [70] and references therein) or in their origin.

Another substantial caveat regarding discussions of particle acceleration and propagation is the distinction between a kinetic plasma theory and ideal MHD. The plasma trapped and propagating in flare loops has a well-developed non-thermal tail (evidenced by HXR and γ-ray emissions), at least during the impulsive phase, which evolves with time, space, and pitch angle. The cartoons of Fig. 2 do not describe any of this. The flare magnetic structure indeed may be much more complicated (Vlahos, this volume).

Microwave and hard X-ray observations show different aspects of the particle distribution functions during the flare evolution. In brief, electrons must stop (the thick-target model) to produce strong hard X-ray emission;

gyrosynchrotron emission, however, is not an appreciable energy-loss mechanism for the hard X-ray energy range, and can persist as long as the electrons are in a strong magnetic field. Mirroring motion in a coronal trap thus enhances the microwave flux relative to the hard X-ray flux and induces a different behavior for the temporal evolution of HXR and microwave spectral indices. This trapping can be demonstrated directly via microwave radioheliograph data, as shown in Fig. 11 looking at the time profiles of loop-top and foot-point sources at 17 and 34 GHz.

In spite of this progress, it has still proven difficult to relate the two populations of non-thermal electrons in the impulsive phase; the hard X-rays generally sample the spectrum at a few tens of keV in the thick target, whereas the microwaves sample mildly relativistic electrons typically in trapping configurations. Because of the tight time correlation between hard X-ray and microwave bursts, the two populations appear to be closely related. Nevertheless discrepancies continue to be reported (e.g., between the spectral indices deduced from the optically thin part of the gyrosynchrotron spectrum and the HXR emitting electrons [72]). The problem may lie in the simplicity of the model assumptions, even in the interpretation of the microwave spectrum [73, 74]. Another solution may lie in the spectral photon flattening observed in several flares above 500-600 keV (e.g., [75, 76]) which may easily explain why electron spectra deduced from >1 MeV gyrosynchrotron-emitting electrons are flatter than the spectra deduced at tens of keV from X-rays. Simple attempts to relate the spectral slopes of the bremsstrahlung and synchrotron-emitting electrons have indeed shown that the centimeter/millimeter-emitting

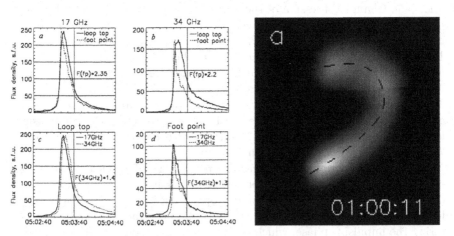

Fig. 11. Left: Clear evidence for trapping, derived from Nobeyama microwave observations, comparing $10'' \times 10''$ photometric boxes to obtain 17 and 34 GHz light curves at footpoint and looptop for for a flare of 13 March 2000 [71]. The delay at the looptop implies trapping. **Right**: Rise phase of a loop flare observed at Nobeyama, showing the clear presence of the footpoint sources at 34 GHz

electrons are related to the hard, high energy region on the HXR/γ-ray spectrum [75]. Such an interpretation could even hold for the combined observations of NoRH and RHESSI of the 23 July 2002 event, given the flattening of the HXR/GR spectrum above 500-600 keV reported by both [77] and [78].

Apart from the number of non-thermal electrons and their low energy cutoffs, microwave observations provide also information on the coronal magnetic field [79, 80]. The exploitation of the precise gyroresonance condition across an active-region corona is exciting future prospect for frequency-agile microwave imaging spectroscopy [79, 81].

What about high-energy ions? Radio techniques do not tell us much directly about their presence, but – depending somewhat upon the abundance of neon [57] – the energy content in ions may even exceed that of the electrons. This further compromises models or theories in the MHD framework, which cannot treat these huge energies self-consistently. MacKinnon & Toner [82] comment on still greater complexities; the energy spectra of accelerated ions may differ by species in flares, as they do when observed in situ.

4 Radio Emission Mechanisms

4.1 Type III Emission from Electron Beams

Type III bursts have been understood conceptually for some decades, but remain an interesting area of theoretical development (E. Kontar & V. Mel'nik, this Workshop). We illustrate this via the data in Fig. 5 (left), which traces an event from GOES soft X-rays out through dekametric and kilometric radio signatures, all the way to the particles observed at the WIND spacecraft near one AU (and the Langmuir waves observed in situ there).

An electron beam, possibly produced in one of the acceleration sites sketched in Fig. 2, runs out along open (type III bursts) or closed (U bursts) magnetic field lines in the corona. As the beam propagates outward, micro-instabilities result in the generation of Langmuir waves, which then couple into the electromagnetic emission that we see. The beam develops with time such that only its leading edge contributes to this process, and we see waves at the plasma frequency or its harmonic from the instantaneous position (hence density) of this moving front. Kontar & Mel'nik (this Workshop) conclude that a gas-dynamic approach can produce the spectrum of Langmuir waves explicitly, and that radio emission can further be calculated using the weak-turbulence approximation. The brightness temperature is found to depend strongly on the beam velocity. Much work remains, though, in the area of obtaining a self-consistent solution in an inhomogeneous plasma that can match the direct observations at 1 AU and also lead to an understanding of the escape of the radio emission.

4.2 Gyrosynchrotron Emissions at Microwaves

Fleishman (this Workshop) presented improved calculations of incoherent and coherent gyrosynchrotron emissions from anisotropic electron distributions. The pitch-angle anisotropy affects the intensity and the spectral index of the emission in the optically thin region, which could explain differences in spectral indices observed for loop-top and foot-point sources with NoRH. Melnikov (this Workshop) also raised the question of Razin suppression in solar centimeter-wave observations (see also [83]). With the Razin cutoff frequency $f_R = 2f_p^2/3f_B \sim 20\ n_e/B$, in cgs units, we can infer values in the range of a few GHz. In the era of full centimeter-wave imaging spectroscopy (i.e., with FASR) it seems likely that this effect will need to be considered.

5 Conclusions

The material covered in Working Group I again demonstrates how broadly significant the radio observations are in our understanding of small scale energy release, acceleration and transport of particles. It is on these small scales (not yet reachable through observations) that particle acceleration can occur. The combination of radio observations, which provide sensitive diagnostics of energetic electrons, with those from other wavelengths provides the best way to study particle acceleration. Our facilities at radio wavelengths (see Table 1) are good but still inadequate in many ways. There seem to be no real technological limits on major improvements in solar radio observations, especially from the point of view of imaging spectroscopy in the cm-mm range where scattering is not so important.

For centimeter-millimeter wavelengths, we look forward to the new facilities (including ALMA and FASR [2]) now being developed. ALMA, for example, will help in understanding the interesting new mm/submm discoveries described here. We believe that FASR will finally *solve* the problem of identifying the site of impulsive-phase particle acceleration, which remains irritatingly uncertain.

At longer wavelengths we look forward eagerly to the STEREO observations, for example, which will for the first time make 3D observations of the motions of coronal sources such as type III bursts or CMEs. The dm-m wavelengths provide the best tools for understanding the dynamics of the middle and upper corona, hard to observe but critically important for the propagation of CMEs and the acceleration of high-energy particles.

Acknowledgments

Hudson was supported by NASA under contract NAS5-98033 with Goddard Space Flight Center.

Appendix

Working group I participants:

D. Burgess, C. Dauphin, B. Dennis, L. Fletcher, G. Fleishman, R. Galloway, P. Giuliani, B. Hamilton, I. Hannah, G. Huang, H. Hudson, M. Karlický, J. Kasparova, E. Kontar, S. Krucker, T. Lüthi, A. Mackinnon, J. Magdalenič, D. Maia, P. Mallik, V. Mel'nikov, T. Neukirch, R. Schroeder, G. Trottet, M. Väänänen, M. Varady, A. Veronig, N. Vilmer, L. Vlahos, P. Wood.

References

1. Klein, K.-L.: Energy Conversion and Particle Acceleration in the Solar Corona. Springer, Lecture Notes in Physics **612** (2003)
2. Gary, D.E., Keller, C.U.: Solar and Space Weather Radiophysics - Current Status and Future Developments. Eds D. E. Gary & C. U. Keller, Kluwer, Astrophysics and Space Science Library, **314** (2004)
3. Gary, D.E., Hurford, G.J.: Coronal temperature, density, and magnetic field maps of a solar acitve region using the Owens Valley Solar Array. ApJ **420** (1994) 903
4. Grechnev, V.V., and 10 co-authors: The Siberian Solar Radio Telescope: the current state of the instrument, observations, and data. Solar Phys **216** (2003) 239
5. Kerdraon, A., Delouis, J.M.: The Nançay Radioheliograph. Coronal Physics from Radio and Space Observations Springer, Lecture Notes in Physics **483** (1997) 192
6. Molodij, G., Trottet, G.: Far Infrared Telescope (TIR) Project. In: Dome C Astronomy and Astrophysics Meeting. Eds. M. Giard, F. Casoli and F. Paletou, EAS Publications Series. (2005) 199
7. Huovelin, J., and 19 co-authors: The SMART-1 X-ray solar monitor (XSM): calibrations for D-CIXS and independent coronal science. PSS **50** (2002) 1345
8. Jain, R., Dave, H., Kumar, S., Deshpande, M.R.: Results of one year of observations of solar flares made by "Solar X-ray Spectrometer (SOXS)" Mission. In: COSPAR, Plenary Meeting (2005) 744
9. Kelley, R.L., and 13 co-authors: The microcalorimeter spectrometer on the ASTRO-E X-ray observatory. Nuclear Instruments and Methods in Physics Research A **444** (2000) 170
10. Hudson, H.S.: Solar flares, microflares, nanoflares, and coronal heating. Solar Phys **133** (1991) 357
11. Parker, E.N.: Nanoflares and the solar X-ray corona. ApJ **330** (1988) 474
12. Hannah, I.G., Christe, S., Krucker, S., Hudson, H.S., Fletcher, L., Hendry, M.A.: RHESSI microflare statistics. In: ESA SP-575: SOHO 15 Coronal Heating. (2004) 259
13. Feldman, U., Doschek, G.A., Behring, W.E., Phillips, K.J.H.: Electron temperature, emission measure, and X-ray flux in A2 to X2 X-ray class solar flares. ApJ **460** (1996) 1034

14. Aschwanden, M.J.: Particle acceleration and kinematics in solar flares - a synthesis of recent observations and theoretical concepts. Space Science Reviews **101** (2002) 1

15. Moore, R.L., Sterling, A.C., Hudson, H.S., Lemen, J.R.: Onset of the magnetic explosion in solar flares and coronal mass ejections. ApJ **552** (2001) 833

16. Fletcher, L., Hudson, H.: The magnetic structure and generation of EUV flare ribbons. Solar Phys **204** (2001) 69

17. Fletcher, L., Pollock, J.A., Potts, H.E.: Tracking of TRACE ultraviolet flare footpoints. Solar Phys **222** (2004) 279

18. Zhang, J., Dere, K.P., Howard, R.A., Kundu, M.R., White, S.M.: On the temporal relationship between coronal mass ejections and flares. ApJ **559** (2001) 452

19. Gallagher, P.T., Dennis, B.R., Krucker, S., Schwartz, R.A., Tolbert, A.K.: RHESSI and TRACE observations of the 21 April 2002 X1.5 flare. Solar Phys **210** (2002) 341

20. Forbes, T.G., Malherbe, J.M.: A shock condensation mechanism for loop prominences. ApJL **302** (1986) L67

21. Masuda, S., Kosugi, T., Hudson, H.S.: A hard X-ray two-ribbon flare observed with *Yohkoh*/HXT. Solar Phys **204** (2001) 55

22. Metcalf, T.R., Alexander, D., Hudson, H.S., Longcope, D.W.: TRACE and *Yohkoh* observations of a white-light flare. ApJ **595** (2003) 483

23. Krucker, S., Hurford, G.J., Lin, R.P.: Hard X-ray source motions in the 2002 July 23 gamma-ray flare. ApJL **595** (2003) L103

24. Svestka, Z.: Solar Flares, Dordrecht: Reidel, 1976 (1976)

25. Sakao, T., Kosugi, T., Masuda, S.: Energy release and particle acceleration in solar flares with respect to flaring magnetic loops. In: ASSL Vol. 229: Observational Plasma Astrophysics: Five Years of *Yohkoh* and Beyond. (1998) 273

26. Qiu, J., Lee, J., Gary, D.E., Wang, H.: Motion of flare footpoint emission and inferred electric field in reconnecting current sheets. ApJ **565** (2002) 1335

27. Asai, A., Yokoyama, T., Shimojo, M., Masuda, S., Kurokawa, H., Shibata, K.: Flare ribbon expansion and energy release rate. ApJ **611** (2004) 557

28. Bogachev, S.A., Somov, B.V., Kosugi, T., Sakao, T.: The motions of the hard X-ray sources in solar flares: images and statistics. ApJ **630** (2005) 561

29. Sui, L., Holman, G.D.: Evidence for the formation of a large-scale current sheet in a solar flare. ApJL **596** (2003) L251

30. Sui, L., Holman, G.D., Dennis, B.R.: Evidence for magnetic reconnection in three homologous solar flares observed by RHESSI. ApJ **612** (2004) 546

31. Somov, B.V.: Cosmic plasma physics. Kluwer Academic Publishers, 2000. Astrophysics and space science library; v. 251 (2000)

32. Veronig, A., et al.: X-ray sources and magnetic reconnection in the X3.9 flare of 2003 November 3. AAp (2006)

33. Forbes, T.G.: A review on the genesis of coronal mass ejections. JGR **105** (2000) 23, 153

34. Hudson, H.S.: Implosions in coronal transients. ApJL **531** (2000) L75

35. Masuda, S., Kosugi, T., Hara, H., Tsuneta, S., Ogawara, Y.: A loop-top hard X-ray source in a compact solar flare as evidence for magnetic reconnection. Nature **371** (1994) 495

36. Lin, R.P., Krucker, S., Hurford, G.J., Smith, D.M., Hudson, H.S., Holman, G.D., Schwartz, R.A., Dennis, B.R., Share, G.H., Murphy, R.J., Emslie, A.G.,

Johns-Krull, C., Vilmer, N.: RHESSI observations of particle acceleration and energy release in an intense solar gamma-ray line flare. ApJL **595** (2003) L69

37. Veronig, A.M., Brown, J.C.: A coronal thick-target interpretation of two hard X-ray loop events. ApJL **603** (2004) L117

38. Wheatland, M.S., Melrose, D.B.: Interpreting *Yohkoh* hard and soft X-ray flare observations. Solar Phys **158** (1995) 283

39. Fletcher, L., Martens, P.C.H.: A model for hard X-ray emission from the top of flaring loops. ApJ **505** (1998) 418

40. Dauphin, C., Vilmer, N., Lüthi, T., Trottet, G., Krucker, S., Magun, A.: Modulations of broad-band radio continua and X-ray emissions in the large X-ray flare on 3 November 2003. Advances in Space Research **35** (2005) 1805

41. Qiu, J., Lee, J., Gary, D.E.: Impulsive and gradual nonthermal emissions in an X-class flare. ApJ **603** (2004) 335

42. Croom, D.L.: Solar millimeter bursts. In: CESRA-2, Committee of European Solar Radio Astronomers. (1971) 85

43. Kaufmann, P., Raulin, J., de Castro, C.G.G., Levato, H., Gary, D.E., Costa, J.E.R., Marun, A., Pereyra, P., Silva, A.V.R., Correia, E.: A new solar burst spectral component emitting only in the terahertz range. ApJL **603** (2004) L121

44. Lüthi, T., Magun, A., Miller, M.: First observation of a solar X-class flare in the submillimeter range with KOSMA. AAp **415** (2004) 1123

45. Hudson, H.S.: The solar-flare infrared continuum - Observational techniques and upper limits. Solar Phys **45** (1975) 69

46. Lüthi, T., Lüdi, A., Magun, A.: Determination of the location and effective angular size of solar flares with a 210 GHz multibeam radiometer. AAp **420** (2004) 361

47. Ohki, K., Hudson, H.S.: The solar-flare infrared continuum. Solar Phys **43** (1975) 405

48. Share, G.H., Murphy, R.J., Smith, D.M., Schwartz, R.A., Lin, R.P.: RHESSI e^+-e^- Annihilation radiation observations: implications for conditions in the flaring solar chromosphere. ApJL **615** (2004) L169

49. Xu, Y., Cao, W., Liu, C., Yang, G., Qiu, J., Jing, J., Denker, C., Wang, H.: Near-infrared observations at 1.56μ of the 2003 October 29 X10 white-light flare. ApJL **607** (2004) L131

50. Lingenfelter, R.E., Ramaty, R.: On the origin of solar flare microwave radio bursts. PSS **15** (1967) 1303

51. Raulin, J.P., Kaufmann, P., Olivieri, R., Correia, E., Makhmutov, V.S., Magun, A.: Time and space distribution of discrete energetic releases in millimeter-wave solar bursts. ApJL **498** (1998) L173

52. Leibacher, J.W., Harvey, J.W., Kopp, G., Hudson, H., GONG Team: Remarkable low-temperature emission of the 4 November 2003 limb flare. American Astronomical Society Meeting **204** (2004)

53. Dauphin, C., Vilmer, N., Krucker, S.: Observations of soft X-ray rising loops associated with a type II and CME emission in the large X-ray flare on 3 November 2003. AAp (2006)

54. Khan, J.I., Vilmer, N., Saint-Hilaire, P., Benz, A.O.: The solar coronal origin of a slowly drifting decimetric-metric pulsation structure. AAp **388** (2002) 363

55. Klein, K.L., Trottet, G.: The Origin of solar energetic particle events: coronal acceleration versus shock wave acceleration. Space Science Reviews **95** (2001) 215

56. Emslie, A.G., Dennis, B.R., Holman, G.D., Hudson, H.S.: Refinements to flare energy estimates: A followup to "Energy partition in two solar flare/CME events" by A.G. Emslie et al. Journal of Geophysical Research (Space Physics) **110** (2005) 11, 103

57. Ramaty, R., Mandzhavidze, N., Kozlovsky, B., Murphy, R.J.: Solar atmospheric abundances and energy content in flare-accelerated Ions from γ-ray spectroscopy. ApJL **455** (1995) L193

58. Klein, K.L., Schwartz, R.A., McTiernan, J.M., Trottet, G., Klassen, A., Lecacheux, A.: An upper limit of the number and energy of electrons accelerated at an extended coronal shock wave. AAp **409** (2003) 317

59. Brown, J.C., Hoyng, P.: Betatron acceleration in a large solar hard X-ray burst. ApJ **200** (1975) 734

60. Karlický, M., Kosugi, T.: Acceleration and heating processes in a collapsing magnetic trap. AAp **419** (2004) 1159

61. Giuliani, P., Neukirch, T., Wood, P.: Particle motion in collapsing magnetic traps in solar flares. I. kinematic theory of collapsing magnetic traps. ApJ **635** (2005) 636

62. Piana, M., Massone, A.M., Kontar, E.P., Emslie, A.G., Brown, J.C., Schwartz, R.A.: Regularized electron flux spectra in the 2002 July 23 solar flare. ApJL **595** (2003) L127

63. Kontar, E.P., Emslie, A.G., Piana, M., Massone, A.M., Brown, J.C.: Determination of electron flux spectra in a solar flare with an augmented regularization method: Application to RHESSI data. Solar Phys **226** (2005) 317

64. Hamilton, B., McClements, K.G., Fletcher, L., Thyagaraja, A.: Field-guided proton acceleration at reconnecting X-points in flares. Solar Phys **214** (2003) 339

65. Hamilton, B., Fletcher, L., McClements, K.G., Thyagaraja, A.: Electron acceleration at reconnecting X-points in solar flares. ApJ **625** (2005) 496

66. Dalla, S., Browning, P.K.: Particle acceleration at a three-dimensional reconnection site in the solar corona. AAp **436** (2005) 1103

67. Efthymiopoulos, C., Gontikakis, C., Anastasiadis, A.: Particle dynamics in 3-D reconnecting current sheets in the solar atmosphere. AAp **443** (2005) 663

68. Zharkova, V.V., Gordovskyy, M.: Particle acceleration asymmetry in a reconnecting nonneutral current sheet. ApJ **604** (2004) 884

69. Hurford, G.J., Schwartz, R.A., Krucker, S., Lin, R.P., Smith, D.M., Vilmer, N.: First γ-ray images of a solar flare. ApJL **595** (2003) L77

70. Miller, J.A., Cargill, P.J., Emslie, A.G., Holman, G.D., Dennis, B.R., LaRosa, T.N., Winglee, R.M., Benka, S.G., Tsuneta, S.: Critical issues for understanding particle acceleration in impulsive solar flares. JGR **102** (1997) 14, 631

71. Mel'nikov, V.F., Reznikova, V.E., Yokoyama, T., Shibasaki, K.: Spectral dynamics of mildly relativistic electrons in extended flaring loops. In: ESA SP-506: Solar Variability: From Core to Outer Frontiers. (2002) 339

72. White, S.M., Krucker, S., Shibasaki, K., Yokoyama, T., Shimojo, M., Kundu, M.R.: Radio and hard X-ray images of high-energy electrons in an X-class solar flare. ApJL **595** (2003) L111

73. Fleishman, G.D., Mel'nikov, V.F.: Optically thick gyrosynchrotron emission from anisotropic electron distributions. ApJ **584** (2003) 1071

74. Fleishman, G.D., Mel'nikov, V.F.: Gyrosynchrotron emission from anisotropic electron distributions. ApJ **587** (2003) 823

75. Trottet, G., Vilmer, N., Barat, C., Benz, A., Magun, A., Kuznetsov, A., Sunyaev, R., Terekhov, O.:   A multiwavelength analysis of an electron-dominated gamma-ray event associated with a disk solar flare. AAp **334** (1998) 1099

76. Vilmer, N., Trottet, G., Barat, C., Schwartz, R.A., Enome, S., Kuznetsov, A., Sunyaev, R., Terekhov, O.: Hard X-ray and γ-ray observations of an electron-dominated event associated with an occulted solar flare. AAp **342** (1999) 575

77. Smith, D.M., Share, G.H., Murphy, R.J., Schwartz, R.A., Shih, A.Y., Lin, R.P.: High-resolution spectroscopy of γ-ray lines from the X-class solar flare of 2002 July 23. ApJL **595** (2003) L81

78. Share, G.H., Murphy, R.J., Skibo, J.G., Smith, D.M., Hudson, H.S., Lin, R.P., Shih, A.Y., Dennis, B.R., Schwartz, R.A., Kozlovsky, B.: High-resolution observation of the solar positron-electron annihilation line. ApJL **595** (2003) L85

79. White, S.M.: Radio measurements of coronal magnetic fields. In: ESA SP-596: Chromospheric and Coronal Magnetic Fields. (2005)

80. Huang, G., Zhou, A., Su, Y., Zhang, J.:   Calculations of the low-cutoff energy of non-thermal electrons in solar microwave and hard X-ray bursts. New Astronomy **10** (2005) 219

81. Bastian, T.S., Benz, A.O., Gary, D.E.: Radio emission from solar flares. ARAA **36** (1998) 131

82. Toner, M.P., MacKinnon, A.L.: Do fast protons and α particles have the same energy distributions in solar flares? Solar Phys **223** (2004) 155

83. Klein, K.L.: Microwave radiation from a dense magneto-active plasma. AAp **183** (1987) 341

Part II

Large-scale Disturbances

Large-scale Waves and Shocks
in the Solar Corona

Alexander Warmuth

Astrophysikalisches Institut Potsdam, An der Sternwarte 16, D-14482 Potsdam,
Germany
awarmuth@aip.de

Abstract. Large-scale waves and shocks in the solar corona are reviewed. The emphasis is on globally propagating wave-like disturbances that are observed in the low corona which have become known as "coronal transient waves" or "coronal Moreton waves". These phenomena have recently come back into focus prompted by the observation of wave-like perturbations in several spectral ranges, particularly in the extreme ultraviolet (with the *SOHO*/EIT instrument). The different observational signatures of coronal waves are discussed with the aim of providing a coherent physical explanation of the phenomena. In addition to imaging observations, radiospectral data are considered in order to point out the relation between coronal waves and metric type II radio bursts. Briefly, potential generation mechanisms of coronal waves are examined. Finally, the relevance of coronal waves to other areas of solar physics is reviewed.

1 Introduction

The solar corona is characterized by a magnetized plasma in which MHD waves and shocks can propagate. It is quite evident that a sudden disturbance of the medium – be it due to a solar flare or an eruption – will launch a wave. The first indications for such globally traveling disturbances were given by the activation of distant filaments by flares, first discussed by Dodson ([22]; see also [80]). Sympathetic flaring (in which a flare seems to trigger another flare in a distant active region) has also been claimed to provide evidence for traveling perturbations (e.g. [9, 104]), but the reality of this phenomenon has remained doubtful (cf. [11]).

Type II solar radio bursts [123], which are seen in dynamic radio spectra as narrow-band emission drifting from higher to lower frequencies, are interpreted as the signature of a collisionless fast-mode MHD shock [98] which expands through the corona and may even penetrate into the interplanetary space (e.g. [13]). The coronal type II bursts are called *metric type II bursts* because they are typically observed at meter wavelengths (for a review, see [4, 58]; see also Gopalswamy, this volume). Using a suitable coronal

A. Warmuth: *Large-scale Waves and Shocks in the Solar Corona*, Lect. Notes Phys. **725**,
107–138 (2007)
DOI 10.1007/978-3-540-71570-2_6

electron density model (e.g. [61, 72]), one can calculate the speed of the type II source, which typically lies around $1\,000$ km s^{-1}.

Large-scale propagating disturbances were finally directly imaged in 1960 using Hα filtergrams [3, 65, 66]. These disturbances, which have since become known as *Moreton waves* or *flare waves*, appear as arc-shaped fronts propagating away from flaring active regions (ARs) at speeds of the order of $1\,000$ km s^{-1}. The fronts are seen in emission in the center and in the blue wing of the Hα line, whereas in the red wing they appear in absorption. This is interpreted as a depression of the chromosphere by an invisible agent [67]. It was also shown that these waves can indeed cause the activation or "winking" of filaments [80].

Uchida [99] developed the theory that Moreton waves are just the "ground track" of a flare-produced fast-mode MHD wavefront which is coronal in nature and sweeps over the chromosphere ("sweeping-skirt hypothesis"). In numerical simulations, Uchida ([100]; see also [101]) was able to show how the waves become focused towards regions of low Alfvén velocity, producing wavefronts that agreed reasonably well with the observations. This model, also known as the blast-wave scenario, can also explain the type II bursts, which are generated at locations where the wavefront steepens to a shock [102]. The association of Moreton waves and type II bursts was also suggested by observations (e.g. [39]).

Since the 1970s, the blast wave scenario has been contested by an alternative model which postulated that coronal shocks are driven by coronal mass ejections (CMEs) acting as a piston (e.g. [17] and references therein). However, this discussion was mainly focused on type II bursts and interplanetary shocks, whereas comparatively little work was done on Moreton waves. This situation was reversed in 1997, when globally propagating wave-like features were detected in the low corona [92] with the Extreme Ultraviolet Imaging Telescope (EIT) aboard the Solar and Heliospheric Observatory (*SOHO*) spacecraft. Since then, wave features have been discovered in several additional spectral ranges. Whether all these signatures are created by the same mechanism is currently intensively debated.

Figure 1 illustrates the different coronal disturbances that can be generated by a solar eruption within the framework of the magnetic reconnection scenario. Reconnection occurs in the diffusion region (DR) below an erupting flux rope (which in this case contains an eruptive prominence – EP). Two pairs of slow-mode standing shocks (SMSS) expand outward from DR, bounding the hot outflowing jets. If the downflow jet is supermagnetosonic a fast-mode standing shock (FMSS; see [6]) is formed above the postflare loops (PFL).

In addition to these standing shocks, propagating waves and/or shocks may be launched. As the erupting flux rope develops into a CME, it can drive a shock provided it is fast enough. This type of shock can reach the outer corona and the heliosphere. The coronal shocks which produce metric type II bursts, on the other hand, may either be launched by the CME or by the flare. At last, there are the large-scale coronal waves which are observed

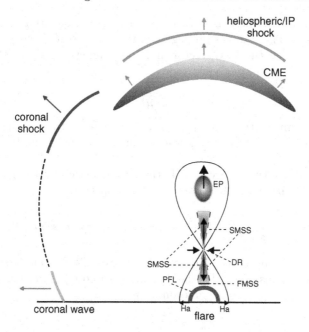

Fig. 1. Schematic representation of the coronal disturbances caused by a solar eruption. For details see main text (adapted from [6])

propagating along the solar surface. They are possibly connected with coronal shocks (indicated by the dashed curve), but it is still far from clear exactly in which manner the different phenomena are related.

In this review, I will focus on the last phenomenon mentioned: the large-scale, globally propagating *coronal waves* (also known as "coronal transient waves" or "coronal Moreton waves"). The basic physics relevant to these phenomena is briefly discussed in Sect. 2. The different observational signatures of the waves are summarized in Sect. 3, while their relation to metric type II bursts is discussed in Sect. 4. Possible physical interpretations of coronal waves are examined in Sect. 5. Potential generation mechanisms of coronal waves are discussed in Sect. 6, and the relevance of coronal waves to other areas of solar physics is reviewed in Sect. 7. The conclusions are given in Sect. 8.

2 The Physics of MHD Waves and Shocks

The solar corona is characterized by a magnetized plasma, which means that disturbances of the medium cannot be treated as purely hydrodynamic. Instead, we have to consider MHD waves and shocks. There are three characteristic MHD wave modes: *Alfvén*, *fast-mode* and *slow-mode* waves. In the case of Alfvén waves, the magnetic tension acts as the restoring force ("shear Alfvén waves"). These waves propagate with $v = v_A \cos\theta_B$, where θ_B is the inclination between the wave vector and the magnetic field, v_A the Alfvén speed

$$v_A = \frac{B}{\sqrt{4\pi\bar{\mu}m_p n}} , \qquad (1)$$

where B is the magnetic field strength, $\bar{\mu}$ the mean molecular weight (taken as $\bar{\mu} = 0.6$ according to [79]), m_p the proton mass, and n the total particle number density.

For fast- and slow-mode waves, both the magnetic and the gas pressure act as restoring forces ("hybrid waves"). Their speed is

$$v_{fm/sm} = (\frac{1}{2}\left\{ v_A^2 + c_s^2 \pm \sqrt{(v_A^2 + c_s^2)^2 - 4v_A^2 c_s^2 \cos^2\theta_B} \right\})^{1/2} , \qquad (2)$$

where c_s is the sound speed. The plus sign gives the fast-mode speed v_{fm}, while using the minus sign yields the slow-mode speed v_{sm}. Another important characteristic speed is the magnetosonic speed

$$v_{ms} = (v_A^2 + c_s^2)^{1/2} \qquad (3)$$

which is the fast-mode speed for $\theta_B = 90°$. For an arbitrary inclination towards B, v_{ms} gives an upper limit for v_{fm}, while v_A or c_s, whichever is greater, is the lower limit (for $\theta_B = 0°$). In many cases v_{ms} is used instead of v_{fm} because θ_B is not known. In the particular case of coronal waves, this is reasonable since they propagate along the solar surface where the magnetic field is predominantly radial.

An important parameter with regard to the propagation of MHD waves and shocks is the ratio of the magnetic pressure to the gas pressure, the so-called plasma beta

$$\beta_p = \frac{8\pi n k_B T}{B^2} = \frac{6c_s^2}{5v_A^2} , \qquad (4)$$

where k_B is the Boltzmann constant and an adiabatic exponent of $\gamma = 5/3$ has been assumed. In most parts of the corona, $\beta_p \ll 1$, which implies also $v_A \gg c_s$. In that case, $v_{ms} = v_A$ can be assumed (i.e., the fast-mode wave has reduced to a compressional Alfvén wave).

So far we have discussed linear waves, which result for linear governing equations. This is an approximation since the basic MHD equations are inherently nonlinear. If a compressive MHD wave has a large amplitude, the nonlinear terms become important and lead to a steepening of the wave's profile. This can be visualized in the following manner: the crest of the wave moves faster than the characteristic velocity of the ambient medium because this speed is locally increased due to the compression. At the same time the leading and trailing edge of the wave still propagate with the ambient characteristic velocity. As a result the wave steepens as shown in Fig. 2. Such

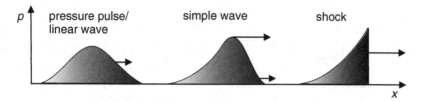

Fig. 2. Schematic of a freely propagating pressure disturbance in the solar corona (pressure p is shown as a function of distance x). An initial pressure pulse (*left*) propagates through the corona as a large-amplitude simple wave (*middle*). The perturbation profile steepens because the wave crest propagates faster than at the leading or trailing edge (*indicated by arrows*). The steepening may lead to the formation of a shock (*right*)

nonlinear large-amplitude waves are called *simple waves* [44, 54]. In the context of this review, we will focus on fast-mode simple waves [59].

Another possibility of a disturbance moving faster than the characteristic velocity of the medium is a *shock wave*. Both fast-mode and slow-mode nonlinear MHD waves can form shocks. A shock is a discontinuity at which the so-called Rankine-Hugoniot or jump conditions have to be fulfilled (see e.g. [79]). Fast-mode and slow-mode shocks are compressive – the downstream density is higher than the upstream one ($\rho_d > \rho_u$). For fast shocks, the downstream magnetic field component parallel to the shock surface increases as compared to the upstream one ($B_d > B_u$), while the converse is true for slow-mode shocks ($B_d < B_u$). Shock speeds can be given in terms of their Mach number, i.e. the Alfvénic Mach number $M_A = v_{shock}/v_A$ or the magnetosonic Mach number $M_{ms} = v_{shock}/v_{ms}$ (note that this nomenclature can also be used for simple waves).

Shocks can also be classified with regard to how they are generated. There are two main types: freely propagating shocks (also called blast-type) and driven shocks. *Freely propagating shocks* start as a large-amplitude disturbance of the medium, which propagates as a non-linear simple wave. The perturbation profile steepens until finally a discontinuity is formed (e.g. [106]) – a shock has been generated (see Fig. 2). As the shock propagates, its amplitude will drop due to geometric expansion, dissipation and the widening of the perturbation profile (the shocked edge moves faster than the trailing one). Ultimately, the shock will decay to an ordinary (i.e. small-amplitude) wave.

In contrast to the blast-type shocks, *driven shocks* are constantly supplied with energy by a driver or piston. There are two subtypes of driven shocks (see Fig. 3) that are often confused. In the true *piston shock* scenario, the medium is confined and cannot stream around the piston. In this geometry, the shock can move faster than the piston, and indeed a shock will be generated even if the piston moves slower than the characteristic speed of the medium. A spherical explosion is another example for such a scenario. In contrast to

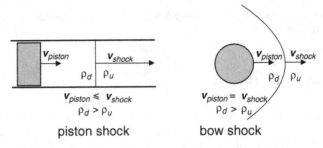

Fig. 3. Schematic of a piston-driven shock (*left*) and a bow shock (*right*)

that, a *bow shock* will form when the medium can stream around and behind the piston. In this case, the shock moves at the same speed as the piston. Moreover, a shock will only form if the piston is faster than the characteristic speed. The best example for this type of shock is the bow shock ahead of Earth's magnetosphere (see Burgess, this volume).

3 Signatures of Coronal Waves

We will now discuss the various observational signatures of coronal waves and their basic characteristics. Unless stated otherwise, the results given in this section are taken from [118].

3.1 Hα (Moreton Waves)

Moreton waves remain the best-studied signature of coronal waves because they have been observed for a long time (e.g. [66]) and because Hα data typically have a much higher time cadence than actual coronal images. A Moreton wave appears as an arc-like front (with an angular width of $\approx 100°$) at some distance from a flaring AR (≈ 100 Mm from the center of the flare). The leading edges of the earliest wavefronts agree very closely with a circular curvature. The front is bright in the line center and in the blue wing of Hα, while it is dark in the red wing. This is interpreted as a depression of the chromosphere by an invisible agent [67]. In the line wings one also sees a fainter front following the first one, where the signature of the intensity change in the wings is reversed as compared to the first front (thus dark in the blue wing and bright in the red wing). This front corresponds to the relaxation of the compressed chromosphere which expands upwards again. The velocity amplitude of the downward swing is some $6\text{-}10 \text{ km s}^{-1}$ [91].

The Doppler shift strongly suggests that the Moreton wave appears only as a reaction to something pressing down from the corona and not due to a wave actually propagating in the chromosphere. This idea is supported by the observed speeds of Moreton waves: for a sample of 15 waves [88] have derived a mean speed of $\langle v_{H\alpha} \rangle = 660$ km s^{-1}, and some Moreton waves reportedly have speeds above $1\,000$ km s^{-1}. However, in the chromosphere the characteristic velocities (e.g. sound speed c_s and Alfvén speed v_A) are of the order of tens of km s^{-1}. If Moreton waves were actually propagating *in* the chromosphere, they would have Mach numbers in excess of 10. Consequently they would suffer strong dissipation and would never propagate over larger distances.

As a Moreton wave moves away from the flare, it becomes increasingly fainter, diffuse and irregular, until its propagation can no longer be tracked. This is the case at distances of ≈ 300 Mm from the flare. An example of the typical evolution is shown in Fig. 4. The signs of decay are also present in the line wings. At the same time, the thickness of the wavefronts becomes larger. All this suggests that the coronal influence to which the chromosphere reacts becomes weaker and less coherent.

A new finding was that Moreton waves are not moving at a constant speed but decelerating [117]. With a sample of 12 events, [118] found an initial Moreton wave speed of $\langle v_1 \rangle = 845 \pm 162$ km s^{-1} (determined from the first wavefront pairs), but an average velocity (obtained from a linear fit) of $\langle \bar{v} \rangle = 643 \pm 179$ km s^{-1}, which agrees closely with the results of [88]. Evidently, Smith & Harvey were using linear fits and did not detect the deceleration. The kinematics of Moreton waves is thus better represented by a 2nd degree polynomial fit. The mean deceleration obtained with this fit, averaged over the 12 waves, is $\langle \bar{a} \rangle = -1\,495 \pm 1\,262$ m s^{-2}. The waves do not display a constant deceleration, instead, the deceleration tends to become weaker with increasing time and distance. For example, if only the first three fronts are used to derive the polynomial fits, then the mean of all decelerations is significantly larger at $\langle \bar{a} \rangle = -2\,460$ m s^{-2}. Thus a power-law fit of the kinematical curves might be even more appropriate. The corresponding power-law index is $\langle \delta \rangle = 0.62 \pm 0.22$.

Moreton waves avoid strong concentrations of magnetic fields, such as ARs. This behavior could be reproduced by the coronal blast wave model of [100], who showed that a coronal fast-mode wave is refracted away from regions of high v_A, i.e. high magnetic field strength.

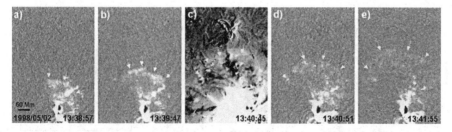

Fig. 4. Evolution of the Moreton wave of 1998 May 2 as shown by Hα (**a, b, d, e**; observed at Kanzelhöhe Solar Observatory) and EIT 195 Å (**c**) difference images. The wavefronts are indicated by *arrows* (from [118])

3.2 Extreme Ultraviolet (EIT Waves)

The Extreme Ultraviolet Imaging Telescope (EIT; see [19]) aboard the Solar and Heliospheric Observatory (*SOHO*) spacecraft has observed globally propagating wave-like disturbances in the corona since 1997 [92, 93]. These so-called *EIT waves* show a wide range of morphological patterns (cf. [49]). Usually they are observed as diffuse and irregular arcs of increased coronal emission in the 195 Å channel of EIT (centered on the Fe XII line, which corresponds to a plasma temperature of ≈ 1.5 MK). Figure 5 shows an example of a strong globally propagating EIT wave.

Limb observations of EIT waves clearly show that they can extend over a significant height range in the low corona, say ≈ 100 Mm. Sometimes EIT waves can be followed across the whole solar disk, which means that they can be tracked to much larger distances than Moreton waves. Note that waves in the EUV have also been observed by the *TRACE* satellite ([124]; see [35], for a description of the instrument).

EIT waves expand away from the site of AR transients (flares, CMEs) at speeds of a few 100 km s^{-1}. This seems to be at odds with the interpretation of EIT waves as the coronal counterpart of Moreton waves, which are on average 2–3 times faster. On the other hand, EIT waves also avoid concentrations of magnetic fields, but they may trigger transverse oscillations of AR loops [124] and filaments [75].

EIT waves are a relatively frequent phenomenon: from 1997 March to 1998 June 173 EIT waves were observed [12]. For comparison, Moreton waves occur roughly an order of magnitude less frequently. Interestingly, about 7% of the events in this big sample display sharp and bright wavefronts somewhat reminiscent of Moreton waves (e.g. [94]) – the so-called "brow waves" [31] or "S-waves" [12]. Such sharp wavefronts are only observed comparatively close to the source AR, and for or several S-waves (cf. Fig. 4) it was shown that they coincide spatially with Moreton waves observed at the same time [47, 118].

This would imply that at least S-waves are the long-sought coronal counterpart to Moreton waves, but what about the more common diffuse EIT waves? Most events showing S-waves also display diffuse fronts at a later

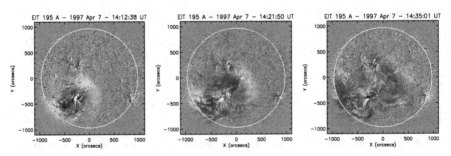

Fig. 5. *SOHO*/EIT 195 Å running difference images showing the globally propagating EIT wave of 1997 April 7

stage, which would be consistent with the idea of a decaying perturbation (see Sect. 3.1). Warmuth et al. [117] have shown that both the sharp and the diffuse EIT wavefronts can be produced by the same disturbance that creates the Moreton waves provided that this disturbance is decelerating (see the distance-time diagram in Fig. 6). The Moreton wave can only be observed relatively near to the source AR where it is still fast. In contrast, the low image cadence of EIT (≈ 15 min) combined with the fact that the waves can be traced to large distances in the EUV means that EIT samples the coronal disturbances when they have already propagated farther away and have thus already decelerated. In this scenario, EIT waves *must* have a lower average speed than Moreton waves. In a systematic study it could be shown that in eight Moreton/EIT wave events this deceleration scenario fits the observations [118]. Other authors (e.g. [23, 75]) have presented events where it is claimed that the two phenomena are distinct. It seems that observations with a higher temporal cadence than EIT will be required in order to positively resolve this issue. Note however that at least deceleration seems to be a characteristic of coronal waves in general, since there are also decelerating EIT waves without associated Moreton waves.

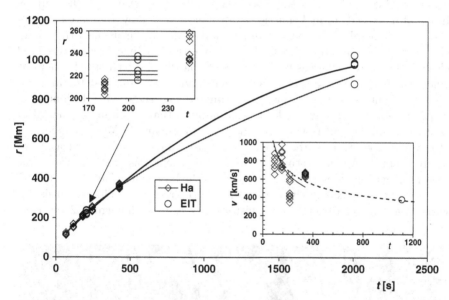

Fig. 6. The combined kinematics of the Moreton (*diamonds*) and EIT wavefronts (*circles*) in the event of 1998 May 2. The main plot shows distance r versus time t. In the upper inset an enlarged part of the graph shows the close association of the Hα and EIT fronts. Error bars are included for the EIT times. 2nd degree polynomial (*thick line*) and power-law (*thin line*) fits are shown. The lower inset shows the velocities of the Moreton wave and the EIT wave. The *thick line* is a fit through the Hα $v(t)$ points, the *thin line* is the derivative of $r(t)$ shown in the main graph (after [117])

Many EIT waves are associated with *coronal dimming* which means a decrease of EUV emission in certain locations in the corona (e.g. [30, 46]). The dimming areas are rather inhomogeneous and can be quite complex. It is now accepted that coronal dimming is usually a result of a mass loss of emitting material, and not primarily due to a temperature change (e.g. [36, 38, 126]). As dimming is generally associated with CMEs, which are in turn often associated with EIT waves, it is well possible that the dimming in wave events is due to an associated CME [129] and not due to the waves themselves.

3.3 Soft X-rays

The observation of coronal waves in the EUV with *SOHO*/EIT came as a surprise since the Soft X-ray Telescope (SXT; see [97]) aboard *Yohkoh* had not observed such phenomena. This can be explained by the observation scheme used by SXT: a flare triggers a special observation mode which minimizes the field of view and the exposure time. These are not favorable conditions for the detection of coronal waves (for details, see [41]).

SXT has finally managed to observe a few coronal waves (see Fig. 7 for an example). Like EIT waves the disturbances observed with SXT show up as fronts of increased coronal emission. Morphologically, they are more homogeneous and generally "sharper" than EIT waves, and in this respect they more closely resemble Moreton waves. This is due to the fact that they are observed close to the source, whereas EIT waves are typically observed only farther out where the disturbance has already weakened and started to disintegrate.

Using a filter-ratio technique, [70] estimated a magnetosonic Mach number of 1.1–1.3 for a SXT wave under the assumption that it is a fast-mode MHD wave. For another SXT wave, [41] derived a comparable Mach number, an electron temperature in the range of 2–4 MK and an emission measure of 5×10^{26} cm^{-5}. An interesting feature of this event was that the wave was seen propagating along the solar limb: it reached a height of up to ≈ 100 Mm and became increasingly tilted towards the solar surface. This is consistent with refraction in a coronal model with v_A increasing with height [62, 99, 120].

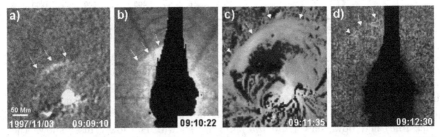

Fig. 7. The SXT wave of 1997 November 3 (**b, d**). The black features in the SXT images are artifacts of saturation. Additionally, the associated Moreton (**a**) and EIT wavefronts (**c**) are shown (from [118])

Since SXT waves are observed relatively close to the source AR, they can be compared to Moreton waves. In all SXT wave events that also had Hα coverage, corresponding Moreton wavefronts were observed (see Fig. 7). It was found that the wavefronts in both wavelength ranges are consistent with a common disturbance [47, 70]. Thus the waves seen in the SXR are really the coronal counterpart to Moreton waves.

Recently, Warmuth et al. [121] have observed global coronal waves with the Solar X-Ray Imager (SXI; see [40, 77]) aboard the *GOES*-12 satellite. Thanks to its cadence (2–4 min) SXI provides a link between the Moreton waves observed close to the AR and the remote EIT fronts. For six events, it could be shown that the wave features seen with SXI are decelerating and agree both with the Hα as well as with the EIT fronts (see Fig. 8 for an example). This is consistent with a single physical disturbance creating all wave signatures.

3.4 Helium I

The Helium I line at 10 830 Å (He I) is formed in a complicated manner (cf. [1]), with influences from the corona, transition region, and chromosphere. Simply put, absorption in the Helium I line increases with increasing UV and EUV flux from the corona and/or with an increase of collisional processes (due to a rise in temperature or density) in the transition region.

Wave signatures were detected in He I [25, 26, 111] with the CHIP instrument [56] at Mauna Loa Solar Observatory. These He I waves are seen in increased absorption. They are more diffuse and thicker than Moreton waves, and have a patchy structure that corresponds with the photospheric magnetic field and the chromospheric network (see Fig. 9 for an example). Some regions behind the He I front show a brightening which coincides with the locations of coronal dimming in cotemporal EIT images [111] observed behind EIT waves.

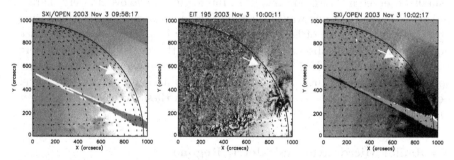

Fig. 8. The propagation of the coronal wave of 2003 November 3 as shown by SXI/OPEN (*left, right*) and EIT 195 Å (*middle*) running difference images. The wave is indicated by *arrows*. Note that the morphology of the wavefront is similar in SXR and EUV. The inclined linear feature in the SXI images is due to overexposure from the flare (from [121])

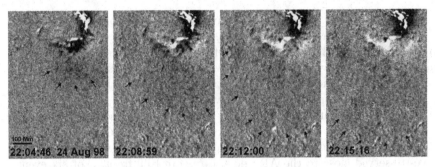

Fig. 9. The He I wave of 1998 August 24 in difference images (*indicated by arrows*). Note its patchy appearance. The flare is seen at the *top* of the images (from [111])

This weakening of absorption in He I is probably due to a reduction of EUV irradiation or heat flux from the corona.

Despite their rather different morphology, He I waves are nevertheless cospatial with both Moreton waves and EIT waves [26, 111]. He I waves also show deceleration, and since they are observed both close to the source AR as well as at larger distances (the temporal cadence is 3 min), they can be regarded as another "missing link" between Moreton and EIT waves (the other one being SXR observations with *GOES*/SXI). Indeed it could be shown for one event that Moreton, EIT and He I wavefronts are consistent with a single decelerating disturbance [111]. Note that despite following similar curves, the He I waves seem to lead the other features by $\approx$ 30 Mm. An analysis of the He I profiles has revealed that they actually have a two-step shape: a shallow perturbation segment ahead of the corresponding Hα front (*forerunner*), and a main perturbation dip which is cospatial with Hα perturbation.

The waves are also visible in He I velocity data (derived from wing observations), where their behavior is consistent with the downward-upward swing usually observed in Moreton waves [27]. Interestingly, two events were characterized by more than one wave – in one of them, five consecutive waves were observed over a period of less than half an hour [27]. This is puzzling since no such behavior was observed in other wavelength ranges. Either these were very special events or He I is more sensitive to wave signatures than other spectral regimes. The authors suggest that the multiplicity of wavefronts may point to more than one generation mechanism in these events, such as flares *and* CMEs (see Sect. 6).

3.5 Radio: Microwaves and Metric Regime

The Nobeyama radioheliograph [68] observes the Sun at 17 and 34 GHz (microwave regime). Aurass et al. [7] first reported a radio feature moving in the same direction as an EIT wave. Warmuth et al. [118] found three events where actual wavefronts were visible at 17 GHz. These fronts, seen as an increase in

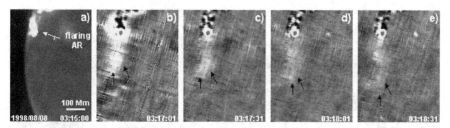

Fig. 10. The coronal wave of 1998 August 8 as shown by 17 GHz difference images (**b–e**). Image (**a**) is a pre-event direct radioheliogram showing the flaring active region and the undisturbed chromosphere (from [118])

microwave emission (see Fig. 10), are cospatial with the associated Moreton wavefronts and are also morphologically similar.

White & Thompson [122] have conducted a detailed study of one of these events. They conclude that the bright wavefronts seem to be more consistent with optically thin thermal free-free emission from the corona than with optically thick chromospheric emission. The observed radio brightness temperatures are consistent with the fluxes of the associated EIT wave if the temperature of the emitting gas is not at the peak formation temperature of the Fe XII 195 Å line or if the abundances are closer to photospheric than to coronal values. The radio brightness temperature declines as the wave propagates, which is consistent with the idea of a disturbance decreasing in amplitude.

Recently Vršnak et al. [115] have discovered wave signatures also in the metric regime (at frequencies between 151 and 327 MHz). With the Nançay radioheliograph [45] they observed a broadband radio source that was moving colaterally with an Hα/ EIT wave. The radio emission is interpreted as optically thin gyrosynchrotron emission excited by the passage of the coronal fast-mode shock.

4 Association with Type II Radio Bursts

Coronal waves which have a large amplitude or which are shocked are potential accelerators of particles. Nonthermal electrons generated in this manner can excite Langmuir turbulence which is subsequently converted to electromagnetic radiation (see [64]). Thus, coronal waves could be sources of type II bursts. Indeed there is observational evidence for this scenario. Smith & Harvey [88] reported that < 50% of Moreton waves were associated with type II bursts, and the comparison of timing and velocities in individual events also suggested a close association between the two phenomena [39, 43].

Recently Warmuth et al. [119] have shown that probably *all* Moreton waves are accompanied by metric type II bursts. The type II bursts in the wave events are ≈ 50% faster and originate lower than an average sample of bursts (for typical type II burst characteristics, see e.g. [16, 60, 84]. This means

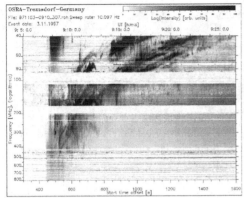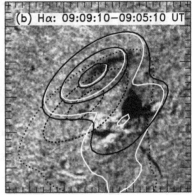

Fig. 11. Left: Metric type II burst associated with a Moreton wave on 1997 November 3 as observed by the Potsdam-Tremsdorf radiospectrograph (see [57]). **Right**: Hα image (Kanzelhöhe Solar Observatory) showing a Moreton wavefront with 30%, 60% and 90% contour levels from a cotemporal Nançay radioheliogram. The type II burst source is cospatial with the Moreton front (from [47])

that they are particularly energetic events. Moreover, close correlations between Moreton and type II kinematics and timing were found, which strongly suggests that Moreton waves and type II bursts are signatures of the same disturbance. This is supported by observations with the Nançay radioheliograph which have shown for two events that the type II burst sources are closely associated with the Moreton wavefronts [47, 78]. One of these examples is shown in Fig. 11.

In four events, [119] measured the band-splitting of the type II emission lanes. Assuming that the band-splitting is due to emission from ahead and behind of the density jump at the coronal shock (e.g. [111]), an Alfvénic Mach number of $M_A \approx 2$ was calculated. Note that this is somewhat higher than the values derived from SXT observations of coronal waves (see Sect. 3.3). It is however consistent with the inferred Mach numbers of the associated Moreton waves [120].

Klassen et al. [49] found that 90% of metric type II bursts are associated with EIT waves. However, the converse is not true: only 21% of EIT waves are accompanied by type II bursts [12]. This suggests that EIT waves are not necessarily associated with coronal shocks, which stands in contrast to the events that do show Moreton wave signatures.

5 The Physical Nature of Coronal Waves

5.1 The MHD Wave/Shock Scenario

We will first discuss the physical nature of the wave events associated with clear chromospheric signatures of Moreton waves because we have a maximum of observational information for these events. In Sect. 3 we have already shown that the waves observed in the different spectral ranges are closely related: Moreton waves are cospatial with sharp EIT waves, SXT and SXI waves, He I waves and waves seen in 17 GHz as well as in metric radioheliograms. Deceleration seems to be a general property of the physical disturbance causing these signatures, which means that also the more remote diffuse EIT waves can be generated by the same perturbation. This is supported by SXI and He I data which bridge the gap between Moreton and EIT wave observations.

The morphology of the signatures (e.g. the fact that they are consisting of enhancements of pre-existing structures) suggests that the common agent is a wave and not, for example, a bulk mass motion like a flare spray. This is strongly supported by the nearly perfect circular curvature of the leading edges of Moreton wavefronts close to their source point [118]. The observed down-up swing of the chromosphere observed in Hα and He I further implies that the impact of a coronal wave leads to the creation of Moreton and He I wave signatures.

Coronal waves are observed over a considerable temperature range (EUV to SXR) and must therefore be compressive, which is independently shown by the microwave data which are sensitive to density enhancements rather than to temperature changes. The waves travel along the solar surface, and since the magnetic field is oriented radially in the quiet Sun, they propagate perpendicularly to the magnetic field. In addition, they are faster than the coronal sound speed (e.g. [62]). These facts suggest that the waves are fast-mode MHD waves (slow-mode waves cannot propagate perpendicular to the magnetic field), and since $\theta_B \approx 90°$ we can treat them as magnetosonic waves.

It should be stressed that the disturbances tend to decelerate to comparable speeds at larger distances, e.g. the mean EIT wave speed given by [118] is $\langle \bar{v}_{EIT} \rangle = 311 \pm 84$ km s^{-1} while a different sample in [121] yielded $\langle \bar{v}_{EIT} \rangle = 320 \pm 120$ km s^{-1}. This implies that in the late phase of the events, the velocities do not reflect the properties connected to an individual event (e.g. the speed of ejected matter in an eruptive scenario), but rather the conditions of the ambient medium (i.e. the magnetosonic speed). This supports the notion that the disturbances are MHD waves.

The magnetosonic speed in the quiet low corona, as given by several authors, is in the range of ≈ 200–600 km s^{-1} [70, 116, 125]. This is consistent with typical EIT wave velocities, but the initial speeds of coronal waves are significantly higher, on the order of $1\,000$ km s^{-1}. This means that at least initially the waves must be shocked, with magnetosonic Mach numbers of up to $M_{ms} \approx 2$–3, as shown for two events in Fig. 12 [120]. Note that also large-amplitude simple waves [59, 106] can move faster than the characteristic speed of the medium. However, a coronal wave can maintain a Mach number of greater than unity over a considerable distance range which means that also its leading edge has to move faster than the characteristic speed. This is only possible for a shock.

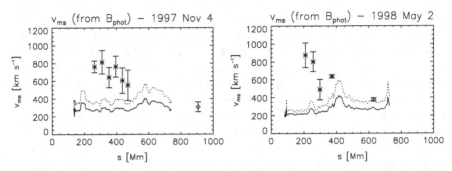

Fig. 12. Comparison of the magnetosonic speed v_{ms} as a function of distance s from the source AR (derived from the measured photospheric magnetic field strength) with measured wave speeds for the events of 1997 November 4 and 1998 May 2. The *solid* and *dotted lines* are the $v_{ms}(s)$ derived for a coronal magnetic field strength of 1/3 and 1/2 times the photospheric value, respectively. *Asterisks* denote Hα Moreton wave speeds, *diamonds* represent EIT wave speeds (after [120])

The shock scenario is supported by the basic characteristics of the waves (deceleration, perturbation broadening and weakening), which are consistent with a shock formed from a large-amplitude simple MHD wave. The simple MHD wave needs time to steepen into a shock which explains both the fact that coronal waves are never observed in the immediate vicinity of their source location. Eventually the shock decays to a linear (i.e. small-amplitude) fast-mode wave (cf. Sect. 2).

An independent confirmation of the shock scenario comes from *Yohkoh/* SXT observations of coronal waves. Using filter ratio methods, Narukage et al. [70] and Hudson et al. [41] have shown that the intensities of coronal waves are consistent with fast-mode shocks. Furthermore, it seems that all Moreton waves are accompanied by metric type II burst, another evidence for a coronal shock. Correlations between the kinematics and timing of Moreton waves and radio bursts, as well as direct comparisons of the locations of wavefronts and burst sources, suggests that coronal waves and type II bursts can be attributed to the same coronal shock.

Figure 13 shows how the different observational signatures are created in this scenario. The curves below the main graph show idealized intensity profiles of the waves seen in Hα line center and HeI (upper plot), the Doppler velocity profile (middle) and the profiles in the wings of Hα (lowermost plot). The variable r denotes the distance from the origin of the wave. Since the coronal magnetosonic speed increases with height in the low corona [62], the shock front is slightly inclined to the magnetic field lines. This is actually observed in limb events [41] and reproduced by numerical simulations [125]. The tilting is also consistent with the "premature" filament activation reported by Eto et al. [23]. The filament (F in Fig. 13) is located higher up in the corona, and is thus activated before the lower parts of the shock have actually

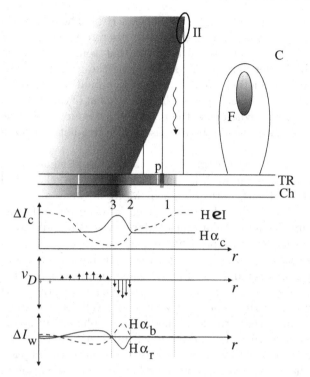

Fig. 13. Schematic presentation of the fast-mode MHD shock passage through the corona (*C*) and its signatures in the transition region (*TR*) and chromosphere (*Ch*). For details see main text (from [119])

reached it. On the disk, the visible EIT wavefront coincides with the lower part of the shock, since the largest fraction of the observed emission is generated there, and the comparatively tenuous upper parts of the wavefront are only observable in limb events.

The downstream coronal plasma is compressed and heated by the shock, creating the wavefronts seen in EUV and SXR. The chromospheric plasma is pushed down by the pressure increase at the coronal base (at $r = 2$), which is observed as the Moreton wave in both Hα line center and wings, Should compressive heating also be taking place in the chromosphere, this would show up as optically thick microwave emission. Alternatively, the compression of the coronal plasma may generate wavefronts in the microwave regime via optically thin free-free emission [122]. After being pushed down the chromosphere relaxes, creating the trailing wavefronts seen in the wings of Hα ($r = 3$).

The pressure jump at and behind the shock causes an increase in density and temperature in the transition region as well. The enhancement of collisional processes could create the main perturbation segment in He I ($r = 2$ in Fig. 13). The He I forerunner ($r = 1$) indicates that processes are influencing the He I absorption already before the shock arrival. The observation that the He I absorption is particularly increased in discrete patches associated with magnetic field concentrations (p) suggests that some agent is propagating down from the higher parts of the shock along the field lines. This could be due to thermal conduction from the shocked coronal plasma or due to electrons accelerated at the quasi-perpendicular segment of the shock (wavy arrow in Fig. 13; see [111]), where also the type II burst source is located (*II*). An alternative explanation would be increased EUV irradiation from the coronal plasma.

The presented model integrates all observational signatures of coronal waves. However, thus far we have only focused on the relatively few events that are associated with both prominent Moreton waves and metric type II bursts. After all, EIT waves have a frequency of occurrence that is about one order of magnitude larger than that of Moreton waves, and only 21% of EIT waves are associated with type II bursts [12].

It may be that in most coronal waves the perturbation is weaker than in the events we have considered. If the wave does not steepen to large amplitudes, or to a shock, no electrons will be accelerated, and consequently no type II burst will be observed. At the same time, a comparatively weak coronal wave will have difficulty perturbing the more inert chromosphere, and no or only weak Moreton wave signatures will be observed. Filaments and coronal loops, on the other hand, appear to be more susceptible towards the impact of coronal waves since they are often excited to oscillate without direct observations of coronal waves [42, 88].

5.2 Alternative Scenarios: Magnetic Reconfiguration

There are of course alternatives to the MHD wave/shock scenario presented above, which are particularly attractive for the weak events (i.e. events without Hα signatures and type II bursts) discussed at the end of the previous section. Inconsistencies between the wave interpretation and EIT observations have first been pointed out by Delannée & Aulanier [20], who noted that in some EIT wave events parts of bright fronts can remain stationary for a prolonged time. Moreover they noted that the bright fronts are followed by an expanding area of coronal dimming. These findings led Delannée & Aulanier [20] to argue that EIT waves are not MHD waves, but rather the consequence of the reconfiguration of magnetic field lines during a CME lift-off.

In this scenario, the stationary bright fronts are due to the compression of the plasma near the footpoints of opening magnetic field lines located close to the separatrix. Dependent on the magnetic topology, such a front might also be propagating as the field opens up further and further from the CME launch site. Such a propagation can be halted when the bright front encounters regions of more or less vertical fields, such as coronal holes or the footpoints of large loops. This was also observed in several cases [21]. Another possibility is that the moving fronts are produced by the interaction of sheared expanding magnetic field lines with surrounding field lines that are nearly potential [21]. This interaction could produce local electric currents, leading to heating that could account for the emission increase in the bright fronts.

Another implication of this scenario is that due to the expansion of the magnetic field lines behind the bright fronts the local plasma density is decreased, which can account for the coronal dimming. Lastly, the fact that CME eruptions often involve large-scale structures such as transequatorial interconnecting loops (TILs) accounts for the observation that many EIT waves are propagating anisotropically, in contrast to such textbook events like the 1997 May 12 event reported by Thompson et al. [92].

In order to obtain more quantitative results (i.e. with respect to propagation velocities), Chen et al. [14] made a numerical simulation of an erupting flux rope and looked for CME-induced wave phenomena. They found that the erupting flux rope drives a piston shock in front of it (see Fig. 14). While the top of this CME-driven coronal shock generates the type II radio burst, its flanks extend down to the solar surface where they can produce Moreton waves (at low altitudes, the shock may degenerate to a finite-amplitude MHD wave). Simultaneously, behind the flanks of the shock a plasma density enhancement is propagating at a lower speed. This feature is due to successive stretching or opening of closed field lines covering the erupting flux rope, and based on its velocity it is interpreted as the EIT wave, which is thus not actually a wave in the physical sense. Corresponding to the plasma enhancement in the EIT front, plasma gets evacuated in the inner region behind the front. This can explain the dimming commonly observed behind EIT waves.

Chen et al. [15] have extended this model by considering a case where two smaller ARs are placed on either side of the erupting flux rope. They found that the density enhancement interpreted as the EIT front stops at the boundary of active regions and coronal holes, which nicely reproduces the behavior of stationary EIT fronts.

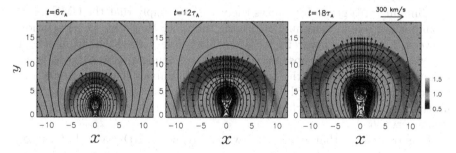

Fig. 14. Evolution of the density (*gray scale*), magnetic field (*solid lines*), and velocity (*arrows*) in the scenario by Chen et al. A piston-driven shock straddles the erupting flux rope, with the flanks sweeping the solar surface and an expanding dimming region lagging far behind (from [15])

The Chen model could be considered as a hybrid between the wave/shock scenario and Delannée's proposal. It explains the velocity discrepancy between Moreton and EIT waves by invoking two physically distinct disturbances, and succeeds in reproducing many observational findings, i.e. that the type II speed is correlated with the Moreton speed, but not with the EIT wave speed. However, it fails in a crucial point: in the well-studied events that have prominent Moreton waves there is no observational evidence for two distinct disturbances. On the contrary, observations linking Hα and EIT observations have clearly shown that a single decelerating disturbance is responsible for both the fast Moreton and the slow EIT wave. Chen et al. [15] argue that in those cases the EIT fronts that are observed are actually the coronal Moreton wave, whereas the predicted slower perturbation is below the observational threshold. In fact, this would fully confirm the wave/shock scenario presented in Sect. 5.1, with the addition of a slow trailing disturbance that seems to be energetically insignificant in most events. In events without Moreton signatures this slow disturbance would nicely reproduce some characteristics of EIT waves (e.g. partly stationary fronts), but why would the perturbation clearly show up in EIT only in those events and not also in the presumably more energetic Moreton-associated events?

Leaving these questions aside, the observation of stationary bright fronts and dimming is actually not in itself sufficient to rule out the wave/shock scenario. In principle, also waves and shocks can trigger localized energy release when they cross pre-existing coronal structures, leading to localized heating and a stationary emission enhancement (cf. [73]). This was possibly the case for a coronal wave observed in the metric regime by Vršnak et al. [115]. When the wave passed enhanced coronal structures, the radio emission became prolonged, indicating that a local energy release was triggered by the disturbance.

The wave/shock scenario does not claim to explain dimming, but it also does not preclude the launch of a CME (which is indeed observed in many events) which would lead to coronal dimming. The dimming could therefore be connected to the CME and not to the wave itself. The observation that the dimming area generally follows the wave may imply that the CME plays an important role in launching or driving the wave. However, the association between dimming and wavefronts is not always very close. For example, in the event studied by Thompson et al. [94], the dimming area was located only behind the eastern half of the bright front. This implies that wavefronts and dimming need not necessarily be as tightly related as the magnetic reconfiguration scenario predicts, where the dimming should always follow the bright fronts very closely.

It may well be that many EIT waves are not MHD waves but rather signatures of a restructuring of coronal magnetic fields, e.g. in the framework of an eruption (e.g. [24, 37]). In particular, this seems to be a fitting scenario for EIT waves that have a very irregular shape or which are very slow or show some kind of erratic propagation. The other extreme of the event spectrum – coronal waves that are associated with prominent Hα signatures and metric type II bursts – shows quite different characteristics that are better reproduced by the wave/shock scenario (Sect. 5.1). In particular, it should be pointed out that all wave signatures in these events are created by a common disturbance that is closely related with the associated metric type II burst. Since type II bursts can only be generated by an MHD shock (and not by magnetic reconfiguration), this implies that also the propagating fronts are created by an MHD shock or wave.

6 Causes of Coronal Waves

While a lot of progress has recently been made regarding the physical nature of coronal waves, their actual causes remain elusive. The same is true for metric type II bursts. It is quite clear that a sudden disturbance has to be introduced into the corona in order to launch large-scale waves or shocks, but there are several candidates for the initial perturbation. Usually, a "flare-driven" and a "CME-driven" scenario are discussed, but the situation is actually more complex than this.

Historically, Moreton waves were first linked to *solar flares* (hence the term *flare waves*) since they are always associated with them. It was noted that the flares in Moreton events were characterized by an "explosive phase" characterized by a sudden increase in brightness and a rapid expansion of the flare borders during the impulsive phase (e.g. [3]). This led to the classical pressure-pulse model where the rapidly expanding flaring volume (effectively acting as a spherical piston; cf. Sect. 2) launches a freely propagating blast wave (see [107]).

Alternatively, *small-scale ejecta* have been proposed as possible causes of coronal waves and shocks. This is based on their speeds which can be comparable to typical Moreton wave speeds (e.g. [108]), as well as on the fact that they are often associated with coronal waves. Flare sprays, observed in Hα, are present in many Moreton wave events [91, 130]. More recently, additional types of flare ejecta have been observed with *Yohkoh*/SXT, such as X-ray jets, plasmoids and erupting loops (e.g. [74, 87]). There is evidence for the generation of metric type II bursts by rapidly expanding X-ray structures [28, 32, 50, 51]. Since these bursts are closely associated with coronal waves, the waves could possibly be launched in the same manner. Physically, such an ejection would act as a temporary piston, generating an initially driven shock. After the ejection stops or decelerates, the disturbance continues as a freely propagating blast wave. Thus in the later stages, there is no difference between the pressure-pulse and the ejection scenario.

The discovery of *coronal mass ejections* in the 1970s led to the "piston-driven" theory of type II bursts (e.g. [17, 33, 89], and references therein). In this scenario a CME acts as a piston creating a driven shock, which can result in a type II burst and/or in a coronal wave.

With regard to the cause of type II bursts, no consensus has yet been reached. There is evidence that both flares *and* CMEs can create shocks (e.g. [16, 86]), but it seems that the flare-generated disturbances usually cannot penetrate to IP space, since most of those bursts cease at $\simeq$ 20 MHz [29]. This is probably due to a local maximum of the Alfvén speed in the higher corona [63]). Therefore, most hectometric/kilometric type II bursts seem to be generated by CME-driven shocks (e.g. [13, 34]).

Let us consider the more complicated situation in the corona in more detail. An excellent timing association of metric type II bursts with the impulsive phase was found (e.g. [48, 90, 105, 109]). Unfortunately this is actually an ambiguous result since the CME acceleration phase is often synchronized with the impulsive energy release of the associated flare (e.g. [113, 127, 128]). Another approach is to look for correlations between various wave/shock characteristics and the flare energy release or CME characteristics. A range of relatively well defined correlations was found for flares (e.g. [76, 109, 110]). Analoguous correlations with CME parameters are either absent or have a low statistical significance [82] unless long-wavelength bursts are also considered. Based on these results, one might suppose that coronal type II bursts are mainly launched by flares. However, since there are type II bursts that extend from the metric regime up to hectometric-kilometric regime (Gopalswamy, this volume), a certain fraction of coronal shocks is probably created by CMEs.

If CMEs are able to create type II bursts in the low corona, they may also be responsible for (some) coronal waves. The most straightforward possibility is that they drive coronal shocks which show up as wavefronts near the solar surface. Another possibility is that they generate fronts that are not due to MHD waves but rather the result of opening magnetic field lines (cf. the discussion in Sect. 5.2). Irregular EIT waves without associated Moreton waves and type II bursts are possibly generated in the latter manner, while the "strong" wave events – with sharp circular Moreton wavefronts, possibly sharp EIT and SXR fronts and type II bursts – seem to be more consistent with a real MHD wave/shock. A third possibility is that the launch of a CME generates an initial pressure pulse which quickly becomes a freely propagating blast wave, much in the same manner as in the flare and small-scale ejection scenarios (e.g. [27]).

How can we distinguish between the different possibilities? An obvious starting point would be to see which associated phenomena are present in coronal wave events. Unfortunately, in the events with prominent Moreton waves flares, small-scale ejecta and CMEs all seem to be present. For EIT waves in general, several authors have used statistical arguments to show that CMEs are a more important ingredient for the production of EIT waves than flares are [12, 20]. At the other end of the event spectrum there are coronal waves which are associated with neither of the potential causes (e.g. [121]).

Researchers are just now beginning to investigate the important issue of the waves' origin, but let us consider some preliminary results anyway. One would expect that a wave caused by a CME, which is a large-scale phenomenon, does not originate from a "point source" such as a flare but from a comparatively extended area. There are reports of flare-associated CMEs which originate from structures with comparatively small sizes, such as the events reported by Neupert et al. ([71]; see also [5]). Still, the preexisting AR loops which were identified as the source of the later CME loops had dimensions of 100–250 Mm. The Moreton waves of Warmuth et al. [118] were first observed ≈ 100 Mm from the source point, where they had a nearly perfect circular curvature and were very sharp. It is difficult to imagine how an extended source such as a CME, even such an initially "compact" CME, could create such signatures. On the other hand, the source point of the waves generally seems to be displaced from the flare center, which is inconsistent with a simple point-like explosion. One might speculate that strong magnetic fields in the active region could provide a guiding of the wave until the outskirts of the AR are reached, where the wave starts to spread out (Huygens' principle). Alternatively, fast-small scale ejecta such as flare sprays might account for the offset.

Regarding the possibility of the CME directly driving a shock as a piston, it should be pointed out that CMEs are accelerating in the low corona (e.g. [127]), whereas coronal waves are decelerating. This rules out the possibility that coronal waves are created by a shock driven by the leading edge of a CME (unless the shock quickly becomes freely propagating in the low corona). However, the flanks of a CME remain fixed during much of the later phase of the eruption, which implies that they have to decelerate somewhere. This means that they could in principle drive a shock that is consistent with the observed wavefronts. This is an important issue for further work, since the kinematical behavior of CME flanks is presently not well understood.

These first results are not sufficient to positively identify the waves' generation mechanism. It is interesting to note, though, that coronal waves that are associated with Moreton waves and metric type II bursts are always accompanied by impulsive flares and/or high-velocity small-scale ejecta. Whether these phenomena constitute a necessary ingredient for the waves' generation remains to be determined. In these events the launch of the waves is closely associated with the impulsive phase of the flares [41, 47, 121], just as it was found for metric type II bursts. Again, the possibly close synchronization of the CME acceleration phase with the impulsive phase of the flare does not allow an unambiguous conclusion.

Many coronal disturbances are not associated with flares or type II bursts, and consequently the situation is much less ambiguous for those events. They may be launched by CMEs or they may be consequences of a restructuring of the coronal magnetic fields (cf. Sect. 5.2). It is also possible that more than one process is working in a single event (cf. [27]).

To make further progress regarding the cause of coronal waves, multiwavelength high-cadence observations of the launch of coronal waves, as well as of the associated flares, small-scale ejecta and CMEs will be required. In addition to radio observations and ground- as well as space-based coronagraphs the space missions *TRACE* [35] and *RHESSI* [55] are particularly important for this task. *TRACE* provides high-cadence and high-resolution observations of coronal processes such as ejections, and might be able to resolve the initiation stage of a coronal wave. The hard X-ray observations of *RHESSI*, on the other hand, allow a detailed analysis of flare energetics.

7 Relevance of Coronal Waves to Other Areas of Solar Physics

Apart from being interesting in themselves and providing information on the flare/CME process, coronal waves can be used to illuminate other aspects of solar physics. In the following, three different "applications" are discussed.

7.1 Particle Acceleration

Coronal waves can have a large amplitude, which means that they are either shocks or large-amplitude simple waves. Both kinds of disturbances are able to accelerate particles, thus they may represent an additional source of *solar energetic particles* (SEPs), which are commonly assumed to be generated at CME-driven interplanetary shocks (for a review, see [81]). Since coronal waves are globally propagating, they can provide an explanation for SEP events that are associated with flares that have a large distance from the Earth-connected magnetic field lines in the western hemisphere of the Sun.

Kocharov et al. [52] first reported the observation of a Moreton wave and an associated SEP event. Based on timing arguments, they concluded that electrons as well as protons are promptly accelerated at the Moreton-associated shock. Torsti et al. [95, 96] claimed that the calculated proton release times are close to the times when EIT waves reach the western limb. A large sample of impulsive electron events was studied by Krucker et al. [53]. For 3/4 of the events that were not related to the flare-associated type III bursts, EIT waves were observed. Krucker et al. conclude (using timing and spatial arguments) that at least some of the impulsive electron events are more likely related to the propagating wave than to the flare itself.

Recently Vainio & Khan [103] have considered particle acceleration at a refracting coronal shock, which means a scenario where the shock front becomes tilted towards the solar surface due to the increase of v_A with height in the low corona (cf. Fig. 13). They noted that in this geometry it is possible that the observer at 1 AU is magnetically connected to the downstream region of the shock (see Fig. 15). Diffusive shock acceleration then results in a power-law spectrum of the accelerated ions – a result which is not naturally obtained when the observer is connected to the upstream region of the shock (as it is the case in the classical CME-driven bow shock scenario). Acceleration in such refracting shocks may also provide a preacceleration mechanism for further acceleration CME-driven shocks in large gradual SEP events.

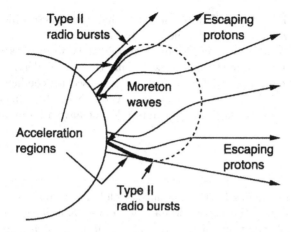

Fig. 15. Global shock geometry of a refracting coronal shock (from [103])

7.2 Coronal Loop Oscillations

The coronal magnetic field is highly elastic, and it is thus expected that it can be excited to oscillate (for theoretical considerations, see e.g. [83]). *TRACE* has indeed observed oscillating coronal loops [2, 69, 85]. These loops oscillate with periods of a few minutes and are damped after a few cycles. This behavior can be explained as a *kink mode* oscillation in which the loop is bodily displaced while the footpoints remain fixed.

The oscillations appear to be excited by nearby flares and filament eruptions. Thus it is well possible that coronal waves launched by the flare/eruption initiate the oscillations. Hudson & Warmuth [42] have supported this scenario through a statistical analysis of 30 oscillation events. In particular, a comparatively high association with metric type II bursts (12 out of 30 events) was found, and the timing of flare/oscillation/type II burst is consistent with the notion that the oscillations are excited by a blast wave associated with the type II burst. This scenario is further supported by the analogy with "winking filaments" and at least one case where a coronal wave is directly observed to excite a loop to oscillate [124].

7.3 Coronal Seismology

Coronal seismology (cf. [83]) is a relatively new diagnostic tool that uses the observed properties of MHD waves and oscillations in order to determine physical parameters of the corona that are otherwise not observable, for example the coronal magnetic field strength. Longitudinal compressive waves in polar plumes [18] and coronal loops [10] and transverse coronal loop oscillations (e.g. [69]) have been used in this manner.

Global properties of the quiet corona, on the other hand, can be derived by studying the propagation of global coronal waves. Mann et al. [62, 63]

have equated the mean EIT wave speed with the magnetosonic speed v_{ms} in the quiet corona. A more detailed study has recently been conducted by Warmuth & Mann [120], where values around 3 G were obtained for the magnetic field strength in the quiet low corona. Ballai & Erdélyi [8] have used the velocity attenuation of EIT waves to derive viscosity coefficients over an order of magnitude higher than the classical value. Note that this approach is only valid if the waves are not shocked and do not have a large amplitude.

8 Conclusions

There has been evidence for the presence of globally propagating, large-scale waves and shocks in the solar corona since more than 50 years. Prompted by spaceborne observations of the corona as well as by high-cadence ground based observations of the chromosphere the recent years have seen a dramatic expansion of our knowledge of these phenomena. Particularly, a large number of global wavelike disturbances has been observed by the *SOHO*/EIT instrument ("EIT waves"). Starting from Hα observations of chromospheric Moreton waves, corresponding wave signatures have been found in the near-IR Helium I line, in the EUV and SXR regime, as well as in radioheliograms. The various signatures all follow closely associated kinematical curves and display deceleration. This implies a common underlying disturbance and resolves the apparent "velocity discrepancy" between Moreton and EIT waves.

The typical characteristics of the common disturbance are deceleration, combined with a broadening of the perturbation profile and a decrease of its amplitude. This is typical for a freely propagating fast-mode MHD shock created by a large-amplitude perturbation (a nonlinear "simple wave"; see [59, 106]). As the shock propagates, its amplitude decreases, which also leads to a deceleration of the disturbance. The presence of a shock is further underlined by the observation of closely associated metric type II radio bursts in all Moreton events. Finally, the shock decays to an ordinary (small-amplitude) fast-mode wave which is supported by the observation that coronal waves decelerate to comparable velocities.

In principle, this scenario is similar to the classical blast-wave model by Uchida [99]. For the "strong events" – those with high initial velocities, clear chromospheric signatures, sharp wavefronts and associated type II bursts – this model fits the observational constraints better than alternative proposals. The wave/shock scenario is particularly supported by the close association found between the coronal waves and metric type II bursts, which can only be generated by an MHD shock. However, it seems that these events form a special class since the majority of coronal waves does not show these characteristics. It may be that in most cases the wave does not steepen to a large amplitude or to a shock, which means that it will become difficult for the wave to perturb the more inert chromosphere. Also, no electrons will be accelerated, and consequently no type II burst will be generated. Alternatively, such "weak events" may not be MHD waves at all in a physical sense, but rather signatures of the restructuring of coronal magnetic fields [15, 21]. This may be a fitting scenario for waves that have an irregular shape, a low speed, or which show some kind of erratic propagation.

At this point a few words have to be said regarding the terminology of coronal waves and shocks. At the moment, there exists a multitude of partly overlapping terms that describe different aspects of these disturbances, and even different physical processes. What is worse is that the usage of terms is somewhat arbitrary and even contradictory at times. A more exact usage of terms is therefore necessary. "Coronal wave" should exclusively be used for moving features that are most likely waves (including large-amplitude simple waves and shocks), such as the "strong events" that are associated with Moreton waves and metric type II bursts. These events actually form a well-defined class: they are all characterized by a wavefront with a relatively smooth shape close to the source AR, a decelerating motion with a mean deceleration of a few 100 $\mathrm{m\,s^{-2}}$ and speeds around 300 $\mathrm{km\,s^{-1}}$ at large distances from the AR.

"Coronal wave" should be considered as the general term referring to the physical disturbance. When there is a need to differentiate these phenomena from other waves in the corona, they can be called "large-scale coronal waves". Terms like "Moreton wave" or "EIT wave" can be used when discussing observations from the respective instruments or spectral ranges. For the multitude of other moving coronal features whose physical nature has yet to be determined, the term "wave" should be avoided altogether. Instead, more general terms such as "coronal transient" or "moving coronal disturbance" should be used. It is true that even the "coronal waves" defined according to the criteria given above are not unambiguously identified as true waves yet. However, as long as there is no convincing evidence to the contrary, the term coronal waves should be retained (though used more discriminately) as it is convenient, has become widely used and reflects the current view of the majority of the solar physics community. A detailed discussion of the issue of terminology can be found in Vršnak [114].

The causes of coronal waves are still unclear. In principle, flares, small-scale ejecta and CMEs are viable mechanisms for the generation of large-amplitude disturbances, while large-scale eruption such as CMEs seem to be the necessary ingredient within the framework of a magnetic reconfiguration scenario. Careful multiwavelength observations of individual events as well as statistical studies will be needed to resolve this issue.

The mere presence of coronal waves signifies that some very impulsive and violent processes must be happening in the early impulsive phase of flares and/or during a CME launch. Coronal waves are thus not only interesting in themselves, but also relevant to other issues in solar physics, including acceleration of solar energetic particles and excitation of loop oscillations. They can even be used to probe the corona for parameters that are otherwise not observable ("coronal seismology"). In order to get a deeper insight into these phenomena, we will need improved observational capabilities which will hopefully be provided by the upcoming missions STEREO and Hinode.

Acknowledgements

I wish to thank the organizers of the CESRA workshop for a particularly stimulating meeting. I also thank G. Mann, H. Aurass and B. Vršnak for fruitful discussions. This work was supported by DLR under grant No. 50 QL 0001.

References

1. Andretta, V., & Jones, H. P.: Astrophys. J. **489**, 375 (1997)
2. Aschwanden, M. J., De Pontieu, B., Schrijver, C. J., & Title, A. M.: Solar Phys. **206**, 99 (2002)
3. Athay, R. G., & Moreton, G. E.: Astron. J. **133**, 935 (1961)
4. Aurass, H.: in *Coronal Physics from Radio and Space Observations*, ed. G. Trottet (Springer, Berlin 1997), Lect. Notes Phys. **483**, 135
5. Aurass, H., Vourlidas, A., Andrews, M. D., Thompson, B. J., Howard, R. H., & Mann, G.: Astrophys. J. **511**, 451 (1999)
6. Aurass, H., Vršnak, B., & Mann, G.: Astron. Astrophys. **384**, 273 (2002a)
7. Aurass, H., Shibasaki, K., & Reiner, M.: Astrophys. J. **567**, 610 (2002b)
8. Ballai, I., & Erdélyi, R.: in *Proceedings of SOHO 13 – Waves, Oscillations and Small-Scale Transient Events in the Solar Atmosphere*, ed. H. Lacoste, ESA SP-547, 433 (2004)
9. Becker, U.: Z. Astrophys. **44**, 243 (1958)
10. Berghmans, D., & Clette, F.: Solar Phys. **186**, 207 (1999)
11. Biesecker, D. A., & Thompson, B. J.: J. Atmos. and Sol.-Terr. Phys. **62**, 1449 (2000)
12. Biesecker, D. A., Myers, D. C., Thompson, B. J., et al.: Astrophys. J. **569**, 1009 (2002)
13. Cane, H. V., Sheeley, N. R., Jr., & Howard, R. A.: J. Geophys. Res. **92**, 9869 (1987)
14. Chen, P. F., Wu, S. T., Shibata, K., & Fang, C.: Astrophys. J. **572**, L99 (2002)
15. Chen, P. F., Fang, C., & Shibata, K.: Astrophys. J. **622**, 1202 (2005)
16. Classen, H. T., & Aurass, H.: Astron. Astrophys. **384**, 1098 (2002)
17. Cliver, E. W., Webb, D. F., & Howard, R. A.: Solar Phys. **187**, 89 (1999)
18. DeForest, C. E., & Gurman, J. B.: Astrophys. J. **501**, L217 (1998)
19. Delaboudinière, J.-P., Artzner, G. E., Brunaud, J., et al.: Solar Phys. **162**, 291 (1995)
20. Delannée, C., & Aulanier, G.: Solar Phys. **190**, 107 (1999)
21. Delannée, C.: Astrophys. J. **545**, 512 (2000)
22. Dodson, H. W.: Astrophys. J. **110**, 382 (1949)
23. Eto, S., Isobe, H., Narukage, N., et al.: PASJ **54**, 481 (2002)
24. Foley, C. R., Harra, L. K., Matthews, S. A., et al.: Astron. Astrophys. **399**, 749 (2003)

25. Gilbert, H. R., Thompson, B. J., Holzer, T. E., Burkepile, J. T.: AGU Fall Meeting 2001, abstract SH12B-0746 (2001)
26. Gilbert, H. R., Thompson, B. J., Holzer, T. E., Burkepile, J. T.: Astrophys. J. **607**, 540 (2004)
27. Gilbert, H. R., & Holzer, T. E.: Astrophys. J. **610**, 572 (2004)
28. Gopalswamy, N., Kundu, M. R., Manoharan, P. K., et al.: Astrophys. J. **486**, 1036 (1997)
29. Gopalswamy, N., Kaiser, M. L., Lepping, R. P., et al.: J. Geophys. Res. **10**, 307 (1998)
30. Gopalswamy, N., & Thompson, B. J.: J. Atmos. and Sol.-Terr. Phys. **62**, 1457 (2000)
31. Gopalswamy, N., Kaiser, M. L., Sato, J., & Pick, M.: in *High Energy Solar Physics – Anticipating HESSI*, eds. R. Ramaty & N. Mandzhavidze (Astronomical Society of the Pacifc, San Francisco 2000), ASP Conf. Ser. **206**, 351
32. Gopalswamy, N., St. Cyr, O. C., Kaiser, M. L., & Yashiro, S.: Solar Phys. **203**, 149 (2001)
33. Gosling, J. T., Hildner, E., MacQueen, R. M., et al.: Solar Phys. **48**, 389 (1976)
34. Gosling, J. T.: J. Geophys. Res. **98**, 18937 (1993)
35. Handy, B. N., Acton, L. W., Kankelborg, C. C., et al.: Solar Phys. **187**, 229 (1999)
36. Harra, L. K., & Sterling, A. C.: Astrophys. J. **561**, L215 (2001)
37. Harra, L. K., & Sterling, A. C.: Astrophys. J. **587**, 429 (2003)
38. Harrison, R. A., Bryans, P., Simnett, G. M., & Lyons, M.: Astron. Astrophys. **400**, 1071 (2003)
39. Harvey, K. L., Martin, S. F., & Riddle, A. C.: Solar Phys. **36**, 151 (1974)
40. Hill, S. M., Pizzo, V. J., Balch, C. C., et al.: Solar Phys. **226**, 255 (2005)
41. Hudson, H. S., Khan, J. I., Lemen, J. R., et al.: Solar Phys. **212**, 121 (2003)
42. Hudson, H. S., & Warmuth, A.: Astrophys. J. **614**, L85 (2004)
43. Kai, K.: Solar Phys. **11**, 310 (1970)
44. Kantrowitz, A. & Petschek, H. E.: in *Plasma in Theory and Application*, ed. W. B. Kunkel (McGraw Hill, Oxford 1966), 148
45. Kerdraon, A., & Delouis, J.-M.: in *Coronal Physics from Radio and Space Observations*, ed. G. Trottet (Springer, Berlin 1997), Lect. Notes Phys. **483**, 192
46. Khan, J. I., & Hudson, H. S.: Geophys. Res. Lett. **27**, 1083 (2000)
47. Khan, J. I., & Aurass, H.: Astron. Astrophys. **383**, 1018 (2002)
48. Klassen, A., Aurass, H., Klein, K.-L., Hofmann, A., & Mann, G.: Astron. Astrophys. **343**, 287 (1999)
49. Klassen, A., Aurass, H., Mann, G., & Thompson, B. J.: Astron. Astrophys. Suppl. **141**, 357 (2000)
50. Klassen, A., Pohjolainen, S., & Klein, K.-L.: Solar Phys. **218**, 197 (2003)
51. Klein, K.-L., Khan, J. I., Vilmer, N., et al.: Astron. Astrophys. **346**, L53 (1999)
52. Kocharov, L. G., Lee, J. W., Zirin, H., et al.: Solar Phys. **155**, 149 (1994)
53. Krucker, S., Larson, D. E., Lin, R. P., & Thompson, B. J.: Astrophys. J. **519**, 864 (1999)
54. Landau, L. D., & Lifshitz, E. M.: *Fluid Mechanics* (Pergamon, Oxford 1987)
55. Lin, R. P., Dennis, B. R., Hurford, G. J., et al.: Solar Phys. **210**, 3 (2002)
56. MacQueen, R. M., Blankner, J. G., Elmore, D. F., et al.: Solar Phys. **182**, 97 (1998)

57. Mann, G., Aurass, H., Voigt, W., & Paschke, J.: in *Proceedings of the 1st SOHO Workshop: Coronal Streamers, Coronal Loops, and Coronal and Solar Wind Composition*, ESA SP-348, 129 (1992)

58. Mann, G.: in *Coronal Magnetic Energy Releases*, eds. A. Benz & A. Krüger (Springer, Berlin 1995a), Lect. Notes Phys. **444**, 183

59. Mann, G.: J. Plasma Phys. **53**, 109 (1995b)

60. Mann, G., Klassen, A., Classen, H.-T., Aurass, H., Scholz, D., MacDowall, R. J., & Stone, R. G.: Astron. Astrophys. Suppl. **119**, 489 (1996)

61. Mann, G., Jansen, F., MacDowall, R. J., et al.: Astron. Astrophys. **348**, 614 (1999a)

62. Mann, G., Aurass, H., Klassen, A., et al.: in *Proceedings of the 8th SOHO Workshop: Plasma Dynamics and Diagnostics in the Solar Transition Region and Corona*, ed. B. Kaldeich-Schürmann, ESA SP-446, 477 (1999b)

63. Mann, G., Klassen, A., Aurass, H., & Classen, H. T.: Astron. Astrophys. **400**, 329 (2003)

64. Melrose, D. B.: Plasma emission mechanisms. In: *Solar Radiophysics*, ed. by McLean, D. J., & Labrum, N. R. (Cambridge Univ. Press, Cambridge 1985), 177

65. Moreton, G. E.: Astron. J. **65**, 494 (1960)

66. Moreton, G. E., & Ramsey, H. E.: PASP **72**, 357 (1960)

67. Moreton, G. E.: Astron. J. **69**, 145 (1964)

68. Nakajima, H., Nishio, M., Enome, S., et al.: Proc. IEEE **82**, 705 (1994)

69. Nakariakov, V. M., Ofman, L., DeLuca, E., et al.: Science **285**, 862 (1999)

70. Narukage, N., Hudson, H. S., Morimoto, T., et al.: Astrophys. J. **572**, L109 (2002)

71. Neupert, W. M., Thompson, B. J., Gurman, J. B., & Plunkett, S. P.: J. Geophys. Res. **106**, 25 215 (2001)

72. Newkirk, G. A.: Astrophys. J. **133**, 983 (1961)

73. Ofman, L., & Thompson, B. J.: Astrophys. J. **574**, 440 (2002)

74. Ohyama, M., & Shibata, K.: PASJ **49**, 249 (1997)

75. Okamoto, T. J., Nakai, H., Keiyama, A., et al.: Astrophys. J. **608**, 1124 (2004)

76. Pearson, D. H., Nelson, R., Kojoian, G., & Seal, J.: Astrophys. J. **336**, 1050 (1989)

77. Pizzo, V. J., Hill, S. M., Balch, C. C., et al.: Solar Phys. **226**, 255 (2005)

78. Pohjolainen, S., Maia, D., Pick, M., et al.: Astrophys. J. **556**, 421 (2001)

79. Priest, E. R.: *Solar Magnetohydrodynamics* (Reidel, Dordrecht 1982)

80. Ramsey, H. E., & Smith, S. F.: Astron. J. **71**, 197 (1966)

81. Reames, D. V.: Space Sci. Rev. **90**, 413 (1999)

82. Reiner, M. J., Kaiser, M. L., Gopalswamy, N., Aurass, H., Mann, G., Vourlidas, A., & Maksimovic, M.: J. Geophys. Res. **106**, 25 279 (2001)

83. Roberts, B.: Solar Phys. **193**, 139 (2000)

84. Robinson, R. D.: Solar Phys. **95**, 343 (1985)

85. Schrijver, C. J., Aschwanden, M. J., & Title, A. M.: Solar Phys. **206**, 69 (2002)

86. Shanmugaraju, A., Moon, Y.-J., Dryer, M., & Umapathy, S.: Solar Phys. **215**, 161 (2003)

87. Shibata, K., Masuda, S., Shimojo, M., et al.: Astrophys. J. **451**, 83L (1995)

88. Smith, S. F., & Harvey, K. L.: in *Physics of the Solar Corona*, ed. C. J. Macris (Reidel, Dordrecht 1971), 156

89. Stewart, R. T., McCabe, M. K., Koomen, M. J., Hansen, R. T., & Dulk, G. A.: Solar Phys. **36**, 203 (1974)

90. Švestka, Z. & Fritzova-Švestkova, L.: Solar Phys. **36**, 417 (1974)
91. Švestka, Z.: *Solar Flares* (Reidel, Dordrecht 1976)
92. Thompson, B. J., Plunkett, S. P., Gurman, J. B., et al.: Geophys. Res. Lett. **25**, 2465 (1998)
93. Thompson, B. J., Gurman, J. B., Neupert, W. M., et al.: Astrophys. J. **517**, L151 (1999)
94. Thompson, B. J., Reynolds, B., Aurass, H., et al.: Solar Phys. **193**, 161 (2000)
95. Torsti, J., Anttila, A., Kocharov, L., et al.: Geophys. Res. Lett. **25**, 2525 (1998)
96. Torsti, J., Kocharov, L., Teittinen, M., et al.: J. Geophys. Res. **104**, 9903 (1999)
97. Tsuneta, S., Acton, L., Bruner, M., et al.: Solar Phys. **136**, 37 (1991)
98. Uchida, Y.: PASJ **12**, 376 (1960)
99. Uchida, Y.: Solar Phys. **4**, 30 (1968)
100. Uchida, Y.: PASJ **22**, 341 (1970)
101. Uchida, Y., Altschuler, M. D., & Newkirk, G., Jr.: Solar Phys. **28**, 495 (1973)
102. Uchida, Y.: Solar Phys. **39**, 431 (1974)
103. Vainio, R., & Khan, J. I.: Astrophys. J. **600**, 451 (2004)
104. Valniček, B.: Bull. Astron. Inst. Czech. **15**, 207 (1964)
105. Vršnak, B., Ruždjak, V., Zlobec, P., & Aurass, H.: Solar Phys. **158**, 331 (1995)
106. Vršnak, B., & Lulić, S.: Solar Phys. **196**, 157 (2000a)
107. Vršnak, B., & Lulić, S.: Solar Phys. **196**, 181 (2000b)
108. Vršnak, B.: J. Geophys. Res. **106**, 25249 (2001a)
109. Vršnak, B.: J. Geophys. Res. **106**, 25291 (2001b)
110. Vršnak, B., Aurass, H., Magdalenić, J., & Gopalswamy, N.: Astron. Astrophys. **377**, 321 (2001)
111. Vršnak, B., Warmuth, A., Brajša, R., & Hanslmeier, A.: Astron. Astrophys. **394**, 299 (2002a)
112. Vršnak, B., Magdalenić, J., Aurass, H., & Mann, G.: Astron. Astrophys. **396**, 673 (2002b)
113. Vršnak, B., Maričić, D., Stanger, A. L., & Veronig, A.: Solar Phys. **225**, 355 (2004)
114. Vršnak, B.: Eos, Vol. 86, No. 11, 112 (2005)
115. Vršnak, B., Magdalenić, J., Temmer, M., Veronig, A., Warmuth, A., Mann, G., Aurass, H., & Otruba, W.: Astrophys. J. **625**, L67 (2005)
116. Wang, Y.-M.: Astrophys. J. **543**, L89 (2000)
117. Warmuth, A., Vršnak, B., Aurass, H., & Hanslmeier, A.: Astrophys. J. **560**, L105 (2001)
118. Warmuth, A., Vršnak, B., Magdalenić, J., et al.: Astron. Astrophys. **418**, 1101 (2004a)
119. Warmuth, A., Vršnak, B., Magdalenić, J., et al.: Astron. Astrophys. **418**, 1117 (2004b)
120. Warmuth, A., & Mann, G.: Astron. Astrophys. **435**, 1123 (2005)
121. Warmuth, A., & Aurass, H., & Mann, G.: Astrophys. J. **626**, L121 (2005)
122. White, S. M., & Thompson, B. J.: Astrophys. J. **620**, L63 (2005)
123. Wild, J. P., & McCready, L. L.: Aust. J. Sci. Res. A **3** 387 (1950)
124. Wills-Davey, M. J., & Thompson, B. J.: Solar Phys. **190**, 467 (1999)
125. Wu, S. T., Zheng, H., Wang, S., et al.: J. Geophys. Res. **106**, 25089 (2001)
126. Zarro, D. M., Sterling, A. C., Thompson, B. J, et al.: Astrophys. J. **520**, L139 (1999)

127. Zhang, J., Dere, K. P., Howard, R. A., et al.: Astrophys. J. **559**, 452 (2001)
128. Zhang, J., Dere, K. P., Howard, R. A., & Vourlidas, A.: Astrophys. J. **604**, 420 (2004)
129. Zhukov, A. N., & Auchère, F.: Astron. Astrophys. **427**, 705 (2004)
130. Zirin, H., & Russo Lackner, D.: Solar Phys. **6**, 86 (1969)

Energetic Particles Related with Coronal and Interplanetary Shocks

N. Gopalswamy

NASA Goddard Space Flight Center, Greenbelt, MD 20771, USA
gopals@ssedmail.gsfc.nasa.gov

Abstract. Acceleration of electrons and ions at the Sun is discussed in the framework of CME-driven shocks. Based on the properties of coronal mass ejections associated with type II bursts at various wavelengths, the possibility of a unified approach to the type II phenomena is suggested. Two aspects of primary importance to shock accelerations are: (1) Energy of the driving CME and (2) the conditions in the medium that supports shock propagation. The high degree of overlap between CMEs associated with large solar energetic particle events and type II bursts occurring at all wavelengths underscores the importance of CME energy in driving shocks far into the interplanetary medium. Presence of preceding CMEs can alter the conditions in the ambient medium, which is shown to influence the intensity of large solar energetic particle events. Both statistical evidence and case studies are presented that underscore the importance of the ambient medium.

1 Introduction

The Sun contributes enormously to the energetic particle population in the heliosphere through various processes: Flares, coronal mass ejections (CMEs), and corotating interaction regions. Electrons are accelerated up to hundreds of MeV; ions are accelerated up to many GeV (see, e.g., [41]). Electrons with energies exceeding 100s of keV produce gyrosynchrotron emission in the microwave to millimeter wavelength range, while the MeV electrons produce gamma-ray continuum. Energetic ions produce very little electromagnetic signatures, except for enhanced nuclear gamma-ray line emissions resulting from the bombardment of the solar surface by accelerated ions. Hard X-ray emission is a common flare signature produced by tens of keV electrons precipitating into the chromosphere from the acceleration site in the corona. Hard X-ray emission is also produced in the corona if the coronal density is high enough to cause thin target bremsstrahlung. Both electrons and ions leaving the Sun are also detected by spacecraft in the solar wind. Very high energy protons occasionally reach Earth's atmosphere, whose effects are detected at the ground level by neutron monitors [36, 43].

N. Gopalswamy: *Energetic Particles Related with Coronal and Interplanetary Shocks*, Lect. Notes Phys. **725**, 139–160 (2007)
DOI 10.1007/978-3-540-71570-2_7

All the nonthermal radio emissions from the Sun are due to accelerated electrons (from a few to 100s of keV energy). Microwave emission during eruptions typically comes from closed loops trapping high energy (100s of keV) electrons. Radio emissions at decimetric and longer wavelengths typically correspond to energetic electrons flowing away from the Sun either from the flare site or from the front of a fast mode MHD shock. Electrons accelerated at shock fronts result in type II radio bursts. Type III radio bursts result from beams of energetic electrons via the beam-plasma instability. Type III bursts occurring without an associated type II are likely to be accelerated at the flare site, but when they occur in association with a type II, the situation is not clear: the electrons may originate in the flare site or at the front of a CME-driven shock (see [3, 54] and references therein). Another related issue is the source of near-relativistic beam-like electrons detected in situ: from the observed time delay of these high-energy electrons with respect to the onset time of the complex type III bursts [27] and the height of the CME at the estimated electron release time, it was concluded that the CME-driven shock is the source of electrons [56].

Ions propagating away from the Sun produce little electromagnetic signatures, so they have to be detected only when they arrive at the observer. Observing the interplanetary type II bursts in association with ions provides information on the shocks that might accelerate the ions. This article focuses on the energetic electrons (that produce type II radio bursts) and ions associated with shocks and CMEs, with a tacit assumption that CME-driven shocks accelerate both electrons and ions. The generation of shocks by CMEs and their propagation through the interplanetary medium are the two primary aspects discussed in this article because they seem to decide the particle output from the shock. The ions discussed in this paper constitute the so-called large solar energetic particle (SEP) events, clearly associated with fast and wide CMEs (see [35, 52] for recent reviews). The large SEP events are also known as gradual SEP events, as opposed to impulsive SEP events, which are of much lower intensity and short-lived. The large SEPs are thought to be accelerated by CME-driven shocks by a shock-drift or diffusive shock acceleration mechanism depending on the geometry of the shock front with respect to the upstream magnetic field. The impulsive particles (see [52]), on the other hand, are thought to be accelerated in the flare reconnection region (see, e.g., [9]). Flares always accompany fast and wide CMEs, so the SEPs may be accelerated in these flares also. How these two processes contribute to the observed interplanetary population of SEPs is not clear. The flare component may be unimportant compared to the shock component [53]. In fact, the largest impulsive event of cycle 23 (April 14, 2001) had an intensity <2 pfu (particle flux unit, 1 pfu = 1 particle per $(cm^2 \, s \, sr)$), five times smaller than the weakest of the gradual SEP events (10 pfu) [21]. The flare component may also be reaccelerated by the associated CME shock to produce charge state and composition different from pure CME events [60].

2 Type II Radio Bursts and Shocks

Type II radio bursts appear as slowly drifting features in the frequency-time plane, a classic representation of radio observations known as the dynamic spectrum. The slow-drift nature was first recognized by Payne-Scott and coworkers [50] at frequencies below 150 MHz and later classified as type II bursts [63]. Extensive literature is available on metric type II bursts that were observed at frequencies above the ionospheric cutoff ($\sim$20 MHz) [48]. Type II bursts in the interplanetary (IP) medium were first detected using spaceborne radio instruments in the early 1970s [44]. The radio emission occurs at the local plasma frequency and its harmonic. Since the local plasma frequency is proportional to the square root of the density, observing type II bursts at various frequencies provides a means of probing various layers of the inner heliosphere. The IP type II bursts also present a strong evidence for particle acceleration far away from the Sun.

The shocks are very strong near the Sun and weaken as they propagate into the heliosphere (see, e.g., [64]), many of them continuing to accelerate particles at 1 AU and beyond. The continued acceleration is inferred from energetic storm particle (ESP) events [51] observed as an enhancement of the particle intensity as the IP shock moves past an observing spacecraft. The shock arrival is also detected as the sudden increase in the frequency of the local plasma line (the low-frequency radio continuum noise) due to the density jump in the shock [28]. Type II bursts observed at frequencies close to the plasma frequency in the vicinity of the spacecraft [1, 38] is another strong evidence for electron acceleration at the location of the observing spacecraft. Whenever the shock is strong enough to accelerate electrons, it can produce a type II burst, irrespective of the distance from the Sun. Figure 1 shows a type II burst observed by the Unified Radio and Plasma Wave (URAP) experiment on board the Ulysses spacecraft. The burst originated from a solar eruption on 2001 May 07 and lasted for three days until the shock arrived (identified as the jump in the local plasma frequency) at the spacecraft on 2001 May 10 (see [57] for details).

The starting frequency of type II bursts is typically below 150 MHz, although higher frequency type II bursts are occasionally observed [61]. The starting frequency is indicative of the distance from the eruption site at which the shock forms. The starting frequency of 150 MHz corresponds to a heliocentric distance of $\sim$1.1 $R_\odot$. Type II bursts observed just above the ionospheric cut-off (15–20 MHz) depart from a distance of $\sim$2.4 $R_\odot$ from the Sun. These distances are based on the Newkirk density model [49] and hence should be considered representative. Thus ground based instruments can observe only the earliest phase of the shocks (propagating over only a small distance of $\sim$1$R_\odot$). Early IP observations were made at frequencies $\leq$2 MHz, so the IP type II bursts were referred to as kilometric (km) because most of the frequency band corresponded to wavelengths in this range. The nominal heliocentric distance of the 2 MHz plasma level is $\sim$20$R_\odot$. The frequency gap

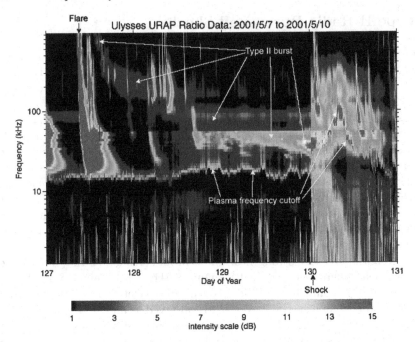

Fig. 1. Dynamic spectrum showing a strong type II burst observed by Ulysses URAP instrument. The start of the solar eruption on 2001 May 07 (DOY 127) and the arrival of the shock on 2001 May 10 (DOY 130) are marked by arrows. Note that the local plasma frequency jumps from ∼12 kHz to >100 kHz at the arrival of the shock, which corresponds to a density jump by a factor of ∼3. See [57] for details. (Courtesy: R. MacDowall)

between 2 and 20 MHz between the IP and metric type II bursts was a source of confusion on the relation between type II bursts in the two domains. The Radio and Plasma Wave (WAVES) experiment on Wind spacecraft [2], essentially filled this gap leading to a significant progress made in understanding the type II bursts.

Information on type II bursts available in various wavelength domains can be combined to produce the schematic in Fig. 2. Each slanted line represents a type II burst: 1) Bursts confined to the metric domain. 2) Bursts starting in the metric domain and ending in the DH domain. 3) Bursts confined to the DH domain. 4) Bursts starting in the DH domain and continuing to the km domain. 5) Bursts starting in the metric domain and having counterparts in all wavelength domains. 6) Bursts found only in the km domain. CMEs associated with the metric type II bursts are of the lowest energy, but are more energetic than the general population of CMEs [40]. On the other hand CMEs associated with DH type II bursts (combining varieties 2, 3, and 4) are more energetic than those associated with the metric type II bursts. The highest energy is possessed by CMEs associated with type II bursts occurring

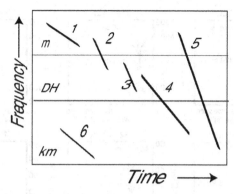

Fig. 2. Schematic dynamic spectra of type II bursts in various wavelength domains (m - metric, DH - decameter-hectometric, km - kilometric). Six varieties of type II bursts can be observationally distinguished on the basis of the wavelength domain in which they occur: 1. confined to the metric domain, 2. starting in the metric domain and ending in the DH domain, 3. confined to the DH domain, 4. starting in the DH domain and ending in the km domain, 5. starting in the metric domain and ending in the km domain, 6. confined to the km domain

at all wavelengths (variety 5). This ordering of CME properties is illustrated in Fig. 3: the speed, width and deceleration of CMEs progressively increase from the general population to those associated with m, DH, and m-to-km type II bursts. This hierarchical relationship between CMEs and type II bursts suggests that the two phenomena are closely related. The purely kilometric type II bursts (variety 6 in Fig. 3) are also associated with CMEs, but they are not as energetic as the all-wavelength type II bursts. Figure 4 shows that the average speed and width of CMEs associated with pure km type II bursts is similar to those of the metric type II CMEs, but the acceleration is positive and much higher (4 m s^{-2}). The positive acceleration implies that the associated CMEs are not super-Alfvenic near the Sun and become super-Alfvenic at large distances from the Sun due to prolonged acceleration. Statistical studies indicate that the slowest CMEs generally show positive acceleration within the LASCO field of view [11, 15]. The propelling force must be acting over large heliocentric distances in these CMEs, which are mostly associated with prominence eruptions. The purely km type II bursts are likely due to shocks that are slightly more energetic than the radio-quiet shocks detected at 1 AU [13, 14]. It must be pointed out that the ISEE 3 type II bursts are likely to be similar to varieties 4 and 5 observed below 2 MHz [4], and not the purely km type II bursts.

Type II bursts is a relatively rare phenomenon [10]: the Solar and Heliospheric Observatory (SOHO) mission's Large Angle and Spectrometric Coronagraph (LASCO) had recorded nearly 7000 CMEs over a period of 7 years (1996–2002), while only 736 metric type II bursts and 350 DH type II bursts were reported over the same period. Thus, not more than ~10% of

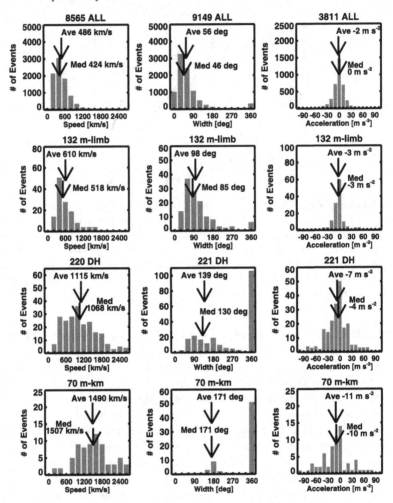

Fig. 3. Speed, width, and acceleration of the four populations of CMEs: the general population (ALL, 1996–2004), the metric type II associated (m-limb), The DH type II associated (DH), and m-to-km type II associated (m-km). The average and median values of the distributions are shown. The average widths were obtained by excluding the full halo CMEs (the last bin in the width histograms). The DH type II bursts represent the combined set, irrespective of the presence of counterparts in other wavelength domains. Type II bursts in the metric domain have their solar sources within 30° from each limb, so that the CME identification is straight forward. Note the progressive shifting of the arrows to the right as one goes from top to bottom rows. In the last column, the arrows shift to the left because the acceleration becomes more negative (deceleration). Note also the increase in the number of full halo CMEs as one goes from the top to bottom in the width column. The average and median widths were computed based on non-halo CMEs

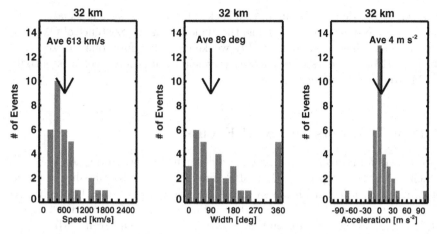

Fig. 4. Speed, width, and acceleration of 32 CMEs associated with purely kilometric type II bursts observed by Wind/WAVES. The average values of the distributions are shown. Note that the acceleration of these CMEs is positive, compared to the other populations shown in Fig. 3

the CMEs are associated with metric type II bursts and not more than ~5% are associated with IP type II bursts. Interestingly, the number of fast (speed ≥ 900 km s^{-1}) CMEs is ~450, similar to the number of DH type II bursts. The numbers do not match exactly because some fast CMEs are not associated with type II bursts while some slower CMEs are associated with type II bursts. Detailed comparison between the occurrence rates of metric and DH type II bursts averaged over Carrington rotation (CR) periods showed much larger discrepancies [10]. For example, CR 1943 had no DH type II bursts, but had a dozen metric type II bursts. Such differences can be explained in terms of the mean speed of the CMEs.

There is a general agreement that the type II varieties 2–6 in Fig. 3 are due to CME-driven shocks. The purely metric type II bursts (variety 1) have been attributed to flare blast waves and CME-driven shocks. The CME-shock possibility was proposed [58] immediately after the discovery of white-light CMEs, but several difficulties arose with the idea stemming from (i) the type II position with respect to the CME, (ii) the relative speeds and directions of the CME and type II, and (iii) CME association. The severest of these problems is the observation of CMEless metric type II bursts [33, 55]: about one third of the metric type II bursts lacked CMEs. These observations became strong pillars supporting the idea that at least some metric type II bursts are of non-CME origin. Reanalysis of the past data which had identified the CMEless metric type II bursts, led to the conclusion that the CMEless type II bursts predominantly originated from the disk center [6]. It is difficult to detect CMEs from the disk center because of the occulting disk employed in coronagraphs to block the direct sunlight. This was also confirmed for a set of disk-center metric type II bursts that lacked CMEs, but were associated with EUV eruptive signatures [14]. Thus, the evidence for the existence of CMEless type II bursts has been considerably weakened. Therefore, one can say that the type II bursts, irrespective of the wavelength of occurrence, is essentially a CME-related phenomenon.

3 Type II Bursts and SEP Events

CMEs were found to be a necessary requirement for the production of SEPs, which led to the suggestion that SEPs are accelerated by the shocks ahead of CMEs [32]. Early studies also indicated a good association between type II radio bursts and SEPs [4, 59]. The utility of Type II bursts in studying SEPs hinges on the assumption that both phenomena result from CME-driven shocks. Type II bursts indicate the departure of shocks near the Sun, so their starting time is likely to be the earliest time SEPs are released. The occurrence rate of large SEP events (proton flux in the > 10 MeV channel ≥ 10 pfu) is similar to that of the DH type II bursts, when averaged over Carrington rotation period; the rate is also not too different from that of shocks detected in situ [10]. This suggests that the same shock accelerates electrons and protons. Furthermore, all (100%) large SEP events of cycle 23 were found to be associated with DH type II bursts, while only 80% were associated with metric type II bursts [19]. For lower energy SEP events, the association rate was lower for DH type II bursts, while the metric type II association rate remained roughly the same [7]. Purely metric events were found to be associated with only small SEP events. The efficiency of the shocks in accelerating particles is likely to increase with heliocentric distance, reaching a maximum above $\sim 3R_\odot$ owing to the shape of the radial profile of the Alfven speed [14]. At these heights, most CMEs would have attained their maximum speeds before decelerating due to coronal drag.

It must be remembered that the occurrence of metric type II bursts is not a sufficient condition for the occurrence of SEP events (see, e.g., [30]). However, when we consider type II bursts occurring at all wavelengths (variety 6 in Fig. 2), the SEP association becomes very high. In fact, the CME properties of m-to-km type II bursts and SEP events are almost identical (compare the last row of Fig. 3 with Fig. 5). The overlap between the two sets of events is ~80% [25]. The lack of 100% overlap can be attributed to the difference in source longitudes: Type II bursts can be observed from CMEs originating from all longitudes (including some slightly behind the limb), while SEPs can be detected only when they have access to the field lines connecting to the observer. The association between type II bursts and SEP events also depends on the efficiency of particle acceleration, which, in turn depends on the CME energy and the properties of the ambient medium.

The strongest observational support to the paradigm that large SEP events are due to CME-driven shocks comes from the correlation between SEP

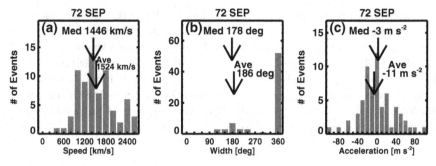

Fig. 5. Speed, width, and acceleration of 72 CMEs associated with large SEP (LSEP) events of cycle 23 (1996–2004). The average and median values of the distributions are shown. Note that the CME properties are nearly identical to those of m-to-km type II bursts (see last row of Fig. 3)

intensity and CME speed [20, 31, 34, 52]. However, the correlation is not perfect, with 3–4 orders of magnitude variation in SEP intensity for a given CME speed. Presence of SEPs in the ambient medium and the spectral variation among SEP events have been proposed as possible factors that could account for one to two orders of magnitude variation in the SEP intensity [31]. Variation in the coronal and IP environment of SEP-producing CMEs may also affect their intensity to a significant extent [20, 23, 24]. SEP intensity may be affected by the presence of turbulence in the vicinity of the shock and in the ambient medium, connectivity of the shock to the observer and the geometry of the shock in the region that is connected to the observer. Most of

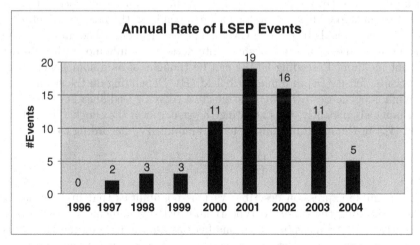

Fig. 6. Annual rate of large SEP (LSEP) events for solar cycle 23 (1996–2004). The actual numbers are indicated at the top of each bin. There was no event in 1996. The largest number (19) was in 2001

these are related to the presence of preceding CMEs in the ambient medium when a powerful shock is launched. The average rate of CMEs near the solar maximum can be as high as 6 per day, so large shocks are likely to encounter preceding CMEs on their path between Sun and Earth. The sharp increase in the number of large SEP (LSEP) events during solar maximum is indicative of such shocks (see Fig. 6). In order to see the effect of preceding CMEs, let us first look at two case studies that illustrate the influence of preceding CMEs on the intensity of large SEP events.

4 CME Interaction

Enhanced radio emission around the time of interaction between two CMEs has been one of the important discoveries of the Wind/WAVES experiment, which was possible due to the simultaneous availability of the radio and white-light observations from SOHO/LASCO [16, 18]. The field of view of SOHO/LASCO (2-32 $R_\odot$) roughly corresponds to the WAVES/RAD2 spectral range (1–14 MHz), which made it easier to visualize the CME interaction at the time of radio enhancement. One of the key results of the interaction studies is that the radio enhancement occurs tens of minutes before the intersection of the leading-edge height-time plots of the two CMEs. This early start is also an indication of the shock ahead of the follower CME penetrating the preceding CME. Most interaction signatures documented so far last for $\leq$ 1h, depending on the size of the preceding CME and the speed of the follower CME. Once the shock traverses the preceding CME, it may merge with the shock of the preceding CME if the latter has one. Beyond that point, the shock propagates through the normal solar wind. Behind the shock is the resultant of the two CMEs, whose final form depends on the relative orientation of the magnetic fields in the two CMEs. The scenario described above can be thought of as an extreme case of shock interaction with inhomogeneities in the ambient medium. Preceding CMEs can be thought of as inhomogeneities in the plasma density (n) and magnetic field (B). Depending on the strength of these inhomogeneities (dB, dn), the effective Alfven speed ahead of the shock can change significantly, thus affecting the strength of the shock. The Alfven speed (V_a) in the medium changes by an amount dV_a according to:

$$\frac{dV_a}{V_a} = \frac{dB}{B} - \frac{1}{2}\frac{dn}{n} \tag{1}$$

This simple expression describes various possibilities that may be useful in understanding the observed type II and SEP phenomena. Coronagraphs are sensitive to the mass in the corona because the underlying mechanism is Thomson scattering by electrons. A preceding CME is a density enhancement ($dn > 0$), so it presents a medium of lower Alfven speed to the follower shock, provided, the magnetic field is not enhanced significantly. This is only an assumption because we do not have direct magnetic field measurements in the corona or CMEs. If $dn/n > 2dB/B$, then the Alfven speed decreases, making the Alfvenic Mach number of the shock $M_a = V_s/V_a$ increase (V_s is the shock speed). The net effect is a stronger shock for the duration of transit through the preceding CME. The shock may not become stronger if the magnetic field is also enhanced. On the other hand if the shock encounters a depletion region ($dn < 0$) the shock can weaken because the local Alfven speed increases. This latter effect may be one of the reasons for the lack of type II radio burst association for a large number of fast and wide CMEs (see later). During the solar maximum phase, the CME rate could be as high as 6 per day, so one can expect fast shocks passing through CMEs of various sizes.

5 CME Interaction and SEPs: Case Studies

We discuss two SEP events in which the unusual SEP intensity can be attributed to the conditions in the corona and IP medium.

5.1 The 2001 April 14-15 Events

Two successive SEP events associated with fast and wide CMEs on 2001 April 14 and 15 occurred from active region 9415 [21]. The weak SEP event of April 14 (SEP1) was associated with an 830 km s^{-1} CME (CME1) and an M1.0 flare; it was the largest impulsive event of cycle 23. The April 15 event (SEP2) was three orders of magnitude more intense ($\sim$1000 pfu) than the April 14th event ($\sim$ 2 pfu) and was associated with a faster CME (CME2, 1200 km s^{-1}) and an X14.4 flare. SEP2 was a large gradual event consistent with the SEP flux (Ip) - soft X-ray flare size (Ix W m^{-2}) relationship [20]:

$$log(Ip) = 4.86 + 0.63log(Ix) \,. \tag{2}$$

According to this relation, the M1.0 flare should result in an Ip of 50 pfu, while the X14.4 flare should be associated with an Ip of 1.2×10^3 pfu. The observed Ip is quite close to the statistical value for the X-flare, but smaller by an order of magnitude for the M flare. On the other hand, the proton intensity and CME speed (V km s^{-1}) are related by [20],

$$log(Ip) = -9.08 + 3.7log(V) \,, \tag{3}$$

which suggests that CME1 and CME2 should have yielded an Ip of 43 and 206 pfu, respectively. Thus, SEP2 had much higher flux than expected from Eq. (3). The eruption configuration itself was quite similar in both CMEs. Both had metric type II bursts, but only the second SEP event had an IP type II burst. The main differences between the two events were: (i) the CME speed (830 vs 1200 km s^{-1}), (ii) connectivity to Earth (W72 vs W84), (iii) associated flare size (M1.0 vs X14.4), (iv) association of IP type II burst (no vs yes) and (iv) the occurrence of preceding CMEs (no vs yes). We suggest that the occurrence of CME1 and the associated flare might have influenced the resulting intensity of SEP2 by providing seed particles and an environment conducive for efficient shock acceleration.

5.2 The 2002 July 20-23 Events

We compare three large eruptions from AR 0039, two on 2002 July 20 and one on July 23 (see [42] for details on this eruption). The region was ∼10° behind the east limb at the time of the July 20 flares while it was at S13E74 at the time of the July 23 flare. We refer to the corresponding CMEs as CME1, CME2, and CME3. Height-time measurements yielded an average speed of 1357, 2017, and 2180 km s^{-1}, respectively for CME1, CME2, and CME3. CME1 had a width of about 73° and did not show substantial lateral expansion. CME2

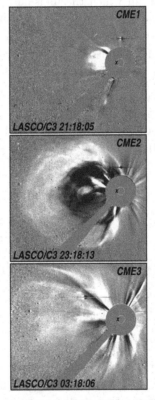

Fig. 7. Difference images of the July 20 and 23 CMEs. **Top:** CME1 (21:18 UT on July 20, 2002), **Middle:** CME2 (23:18 UT on July 20), **Bottom:** CME3 (03:18 UT on July 23, 2002). The C3 occulting disk (*shown as the gray circle in each image*) has a radius of 3.7 R$_\odot$. The northern section of CME2 is bright probably because it has merged with CME1 at this location. The locations of the flares are marked by 'x' close to the east limb in each case

and CME3 expanded considerably and became asymmetric halo events (faint extensions can be seen above the west limb in the middle and bottom images of Fig. 7). CME1 was overtaken by CME2 within the C3 field of view, so we see the merger of CME1 and CME2 in the middle panel of Fig. 7. Note that CME2 and CME3 are very similar in appearance.

Proton intensities as measured in the > 10 and > 30 MeV channels of the ACE/SIS instrument are plotted in Fig. 8. The intensities in both channels were elevated from normal background levels prior to the slow rise following the July 20 CMEs. They both reached a broad peak around 09:00 UT on July 22 and continued at this level past the onset time of the July 23 CME. The rise time was about 37 hours, typical of east limb events. In the higher energy (> 30 MeV) channel, the intensity peaked slightly before that in the > 10 MeV channel and declined monotonically thereafter. On the other hand, the July 23 eruption was not associated with a similar particle increase. There was an impulsive peak on July 24 at 13:00 UT, but it is small and short-lived compared to the main event. There was a small peak around 13:00 UT on July 25, which coincided with an IP shock detected in situ by SOHO, ACE and Wind. This shock is likely to be associated with the July 23 CME. We are confident that the large SEP event was due to CME2 because no other CME of importance occurred during the rising phase of the SEP event. Only two other CMEs were noted between the onset of CME2 and the peak of the SEP event (09:00 UT on July 22): (i) a large blob structure along a streamer in the northeast on July 21 at 02:30 UT, and (ii) a fairly faint loop from the south on July 22 at 09:30 UT. The speeds of these CMEs were not high enough to drive shocks and their locations were not favorable for producing SEP events at the Earth.

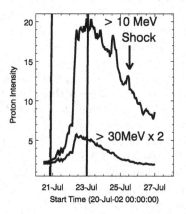

Fig. 8. Intensity of protons in units of $(cm^2 \ s \ sr \ MeV)^{-1}$ in the >10 MeV and >30 MeV channels as marked. The >30 MeV intensities have been multiplied by a factor of 2. The height-time plots (*the vertical lines*) of the three CMEs at the Sun are also shown. Note that the first two CMEs are so close in time that the height-time plots cannot be seen distinctly. The vertical arrow points to the arrival time of the IP shock at ACE from the July 23 CME

Was there an SEP event associated with the July 23 CME? All we see is a small blip at ~12UT on July 24 on the decaying profile of the previous event. By shifting the SEP profile of the July 20 event to the onset time of CME3, one can confirm that this small blip is likely to be the SEP event associated with the July 23 CME. If we subtract the decaying profile of the July 20 SEP event, the intensity level can be inferred to be an order of magnitude smaller. Mars was located directly behind the Sun at the time of the July events (the Earth-Sun-Mars angle was ~169°). The particle observation at Mars was similar to that at Earth: the July 20 CME was associated with a significant SEP event at Mars, while the July 23 event was not (D. Mitchell 2004, private communications). What factors could have been responsible for the lack of SEPs during the July 23 event? When we looked at the coronal and IP environment of the two events, we found a marked difference: The July 20 CME (CME2) was preceded within an hour by CME1 from the same active region (AR 0039) and the two CMEs seem to have interacted within the coronagraph field of view. Furthermore, the July 20 event was preceded by two other big events on July 19, which might have distorted the field lines overlying AR 0039. Thus the coronal environment of the July 20 CME-driven shock was vastly different from that of the July 23 shock.

6 SEP Intensity Variation: Statistical Study

The two case studies presented above suggest that the coronal and IP environment of fast and wide CMEs may play an important role in determining the intensity of the associated SEP events [18, 20]. What emerges from these studies is that the intensities of SEP events may be ordered by the presence of preceding CMEs in the near Sun IP medium. A recent study [23] identified the primary CME and the solar source region for each of the large SEP events during 1996–2002. Based on the occurrence of preceding CMEs within a day ahead of the primary CMEs, the SEP events were grouped as: (1) those with preceding wide CMEs (P events), (2) those without preceding wide CMEs (NP events), and (3) those, which do not belong to P or NP events either because of a possible preceding CME that was overtaken by the primary CME below the coronagraph occulting disk or the primary CME interacted with a nearby streamer (O events). Only preceding CMEs originating from the same active region as the primary CME were considered. The CME and flare properties of the three groups of events are summarized in Table 1. The median intensity (Ip) of the >10 MeV protons is smallest for the NP events and largest for the P events (column 2 in Table 1). However, most of the CME properties of the P and NP events are similar: The speeds are virtually the same, but the SEP intensities differ by an order of magnitude. The masses are similar and hence the kinetic energies are also similar for the P and NP events. Note also the larger flare size and active region area for the P events, compared to the NP events (see Sect. 7).

Table 1. Properties of Eruptions associated with P, NP, and O events

Group	Ip (pfu)	V (km s^{-1})	M (10^{15} g)	KE (10^{31} erg)	Flare Size	AR Area (msh)
ALL	54	1379	9.0	8.5	M7.1	770
P	210	1300	8.1	7.1	X2.1	880
O	91	1385	11.3	8.5	M2.3	620
NP	29	1379	9.6	10.0	M3.9	500

The scatter plot between CME speed and SEP flux is shown in Fig. 9, with the three populations (P, NP, and O) distinguished. We have plotted 30 P, 14 NP, and 13 O events that had solar sources on the frontside of the Sun. The overall correlation coefficient (r) is 0.57, similar to values obtained by others [20, 31]. The O events had the highest correlation (r = 0.74) followed by P events (r = 0.64) and the NP events (r = 0.36). Note that the NP events occupy the bottom portion of the plot area (very low intensity), while the P and O events had higher intensity (see also Table 1). The combined set of O and P events had a correlation of 0.67, which is better than that of the NP events by 54%. The correlation also improved when backside events were excluded. Combining the P and O events, we see that for the majority of SEP

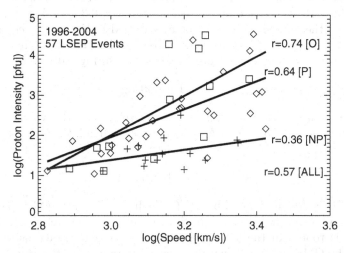

Fig. 9. Scatter plot between SEP intensity (*pfu*) and CME speed (km s^{-1}) in logarithmic units for 57 frontside events (1996–2004). The P (*square*), O (*diamond*) and NP (*triangle*) events are distinguished. Note that most of the NP events had intensity less than 100 pfu, whereas many of the P and O events had intensity exceeding 1000 pfu. The correlation coefficient for all the SEP events is 0.57, while it is higher for the P (r = 0.64) and O (r = 0.74) events. The NP events had the lowest correlation coefficient (r = 0.36). The regression lines for the three groups of events are also shown by the straight lines

events (43/57 or 75%), the primary CMEs propagated through the aftermath of preceding CMEs, and not through the normal solar wind. The scatter is also reduced when the P and NP populations are considered separately, so it can be concluded that the preceding CMEs is a source of scatter in the CME speed – SEP intensity plots (see [31] for other sources of scatter).

7 SEP Intensity and Active Region Area

In Table 1, we saw that the flare size of the P events is higher than that of the NP events. Is it possible that the intensity variation may have something to do with the properties of the source region? One way to look at this is to examine the active region area. Larger active regions can store greater amounts of free energy resulting in larger eruptions. To do this, we selected all the active regions that had at least one large SEP event during disk passage [24]. We obtained the area information (in units of millionths of solar hemisphere – msh) for all the regions from the Solar Geophysical Data. Figure 10(a) shows the distribution of AR areas associated with the SEP events. Most of the regions had areas < 900 msh. The median areas of the P, O, and NP events were not too different (see Table 1).

The X-ray flare size is reasonably correlated with the active region area (correlation coefficient r=0.61, see Fig. 10(b)). There is also a weak correlation between the CME speed and active region area (r = 0.38, see Fig. 10(c)). In other words, the flare size seems to be closely tied to the active region area as compared to the CME speed. The regression line for the flare size (Ix W m^{-2}) – active region area (A msh) scatter plot in Fig. 10(b) is given by,

$$\log Ix = -8.34 + 1.50 \log A .\qquad(4)$$

Interestingly, the X-ray peak flux is linearly related to the volume of the active region, because volume $\sim A^{3/2}$. On the other hand, the CME speed (V km s^{-1}) and A in Fig. 10(c) are related by,

$$\log V = 2.54 + 0.22 \log A .\qquad(5)$$

If the SEPs are accelerated by the CME-driven shock, the SEP intensity need not have a specific relationship with the active region area. Even though the CME is rooted in the active region, the three-dimensional shock front ahead of the CME, when it is within a few solar radii from the Sun, is much larger than the active region area. The SEPs are released from the shock front surrounding the moving volume of the CME. This might explain why there is no relationship between the SEP intensity and active region area. On the other hand, the flare is confined to the arcade of loops in the active region and filled with hot plasma emitting soft X-rays. The flaring loops are generally confined to the active region, so one can understand the good correlation between flare size and active region area. In other words, the flare energy and the SEPs

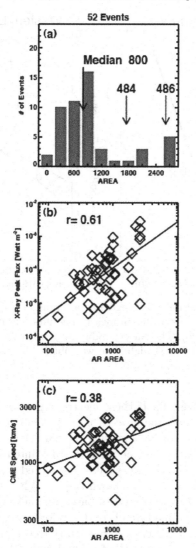

Fig. 10. (a) Distribution of active region areas associated with the LSEP events from 1996–2003. (b) scatter plot between X-ray flare size (measured as the peak intensity in the 1–8 Å GOES X-ray channel) and AR area. (c) scatter plot between CME speed and AR area. The corresponding correlation coefficients and regression lines are shown in (b) and (c). The two largest active regions 486 and 484 that produced some of the most intense events of cycle 23 are indicated in (a)

are released at two different spatial locations, on either side of the moving CME plasma. Yet, the three physical entities (CME, flare loops, and SEPs) seem to be related in a complex way. This result may also be relevant to the previous result that the intensity of energetic (108 keV) electrons are well

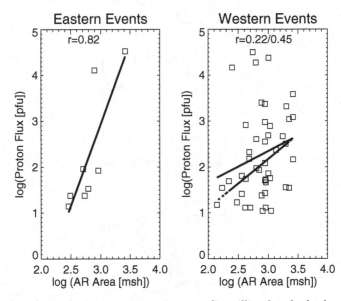

Fig. 11. Correlation between active region area (in millionths of solar hemisphere – msh) and the proton intensity (pfu) in logarithmic units. **Left**: Eastern events (8), **Right**: western events (46). The corresponding regression lines are also shown. For the western events, the lower straight line was obtained when the four outliers of very high intensity are excluded (r = 0.45)

correlated with flare size [10]. If these electrons propagating away from the sun are accelerated by the same process as that of the precipitating electrons, one might expect a good correlation with the flare size.

Figure 11 shows a scatter plot between the SEP intensity and active region area. The two quantities are only weakly correlated (correlation coefficient r = 0.31 for all the events), as was shown in [24]. When the events were separated into eastern and western events based on their source longitude, the correlation coefficient remained small (r = 0.20) for western events, while it became higher (r = 0.82) for the eastern events. However, the sample of eastern events is too small. If we exclude the four high-intensity western events, the correlation coefficient improved to r = 0.45. Thus there is an overall correlation between the SEP intensity and AR area, but weak.

8 Discussion and Conclusions

The possible non-existence of CMEless type II bursts and the association of progressively more energetic CMEs with metric, DH and full-range type II bursts suggests the importance of CMEs for the type II phenomena, and by extension, for the SEP events. There are several related issues that need to be addressed in the framework of CME-driven shocks.

For example, a significant fraction of fast and wide CMEs are not associated with any type II radio bursts above the detection threshold of radio instruments [15]. These radio-quiet CMEs pose a significant challenge to the CME-driven shock hypothesis. Furthermore, a large number of metric type II bursts and DH type II bursts are associated with slow CMEs (speeds < 400 km s^{-1}). This apparent paradox originates from the definition of fast and slow CMEs. Traditionally, CMEs faster than $\sim$400–500 km s^{-1} were thought to be fast because they are super-Alfvenic according to the canonical value of coronal Alfven speed used in the literature (see e.g., [26, 29]). This value may be generally true for regions above the source surface, but it may not be so in the coronal region where metric type II bursts originate. A careful look at the variation of density and magnetic field variation in the inner corona reveals that the Alfven speed could be as low as 200 km s^{-1} near the base of the quiet corona [39] with a peak of several hundred km s^{-1} around 3 R$_\odot$. This fact has been taken to suggest that flare blast waves should have speeds exceeding the Alfven speed peak to become interplanetary shocks [45]. Type II bursts mostly originate from the active region corona, so the Alfven speed profile must be modified by introducing the magnetic field of active regions [14]. The modified profile is similar to the quiet-sun profile except for the rapid increase towards the coronal base to a few thousand km s^{-1} at the core of the active region [14, 46]. Interestingly, CMEs start from rest at very small coronal heights before attaining significant speed. This combination of circumstances might explain the relatively low starting frequency of type II bursts because it is difficult to shock the core of the active region corona. Thus a 300 km s^{-1} CME could be a fast CME while a 1000 km s^{-1} CME could be a slow one depending on the value of the ambient Alfven speed. If this result is combined with the requirement that a CME needs to be present for a type II burst production, one can explain the type II phenomenon solely in terms of CME-driven shocks (see also [8, 25]). Such a conclusion would also be consistent with the idea that SEPs are due to CME-driven shocks. The close association between CMEs and type II bursts suggests that the associated flares need to be eruptive, i.e., they are associated with mass motion. This means the non-eruptive flares, which constitute the majority, are not associated with type II bursts.

While there is universal agreement that CME-driven shocks are responsible for the type II bursts in the interplanetary medium (varieties 2-6 in 2), there is no such agreement on those (variety 1 in 2) confined to the metric domain (see, e.g., [37]). It is often argued that the presence of a CME during a metric type II does not imply a physical relationship between the two phenomena (see, e.g., [62] and references therein). While this might become clear when CME observations extend to lower heights such as by the coronagraphs on board STEREO mission to be flown soon, the fact remains that even X-class flares do not produce type II bursts if they are non-eruptive (i.e., if they lack CMEs). As far as SEPs are concerned, purely metric type II bursts are unimportant.

The Alfven speed profile also provides a natural explanation for the CME interaction phenomenon in terms of a temporary change in the Alfven speed due to the presence of preceding CMEs. The density enhancement in the ambient medium caused by the preceding CMEs can temporarily reduce the Alfven speed, so that the shock entering such a CME will attain a higher Mach number and hence accelerate more particles. More electrons would imply higher intensity radio emission, as has been observed in the case of colliding CMEs. A generalized version of such CME interaction is the propagation of a CME-driven shock through a series of preceding CMEs of various sizes and amplitudes, i.e., enhanced turbulence. The opposite end of the density effect is to have a depletion in the ambient medium, which would result in weakening of the shock and consequently less number of particles would be accelerated. This would mean a weaker radio emission and less number of SEPs. Eruption near a coronal hole presents such a situation, because the flank of the shock nearer to the coronal hole is likely to be weaker than the opposite flank. The effect of the turbulence can also lead to severe distortion of the shock front causing corrugation as occasionally observed [1, 38]. Such an inhomogeneity in shock strength was also visualized early on and referred to as 'Swiss Cheese effect' [5].

One other consequence of the preceding CMEs is that they can accelerate particles to low levels, which may serve as seed particles for the follower shock. Most of the large SEP events are also accompanied by flares, which also accelerate particles. Flare-related impulsive events are of short duration, small in size, but more frequent. The CME-related gradual events are less frequent, but last longer and have much higher intensity. Although there are separate investigations of SEP seed particles [31] and flare seed particles [47] there is no systematic investigation to assess the relative importance of these two populations as seed particles. However, one can speculate that the SEP seed particles may contribute to the intensity of the resulting SEP event, while flare particles may alter the composition of the accelerated particle population.

Acknowledgments

I thank S. Yashiro and E. Aguilar-Roriguez for help with some figures. This material is based upon work supported by the National Aeronautics and Space Administration under the Living with a Star (LWS) and Supporting Research and Technology (SR&T) programs.

References

1. S. D. Bale, M. J. Reiner, J.-L. Bougeret et al.: Geophys. Res. Lett. **26**, 1573 (1999)
2. J.-L. Bougeret, M. L. Kaiser, P. Kellogg et al.: Space Sci. Rev. **71**, 231 (1995)
3. J.-L. Bougeret, P. Zarka, C. Caroubalos et al.: Geophys. Res. Lett. **25**, 2513 (1998)

4. H. V. Cane and R.G. Stone: Astrophys. J. **282**, 339 (1984)
5. J. K. Chao: In *Solar Wind Three*, ed by C. T. Russell (Univ. California Los Angeles, 1974), pp. 169–174
6. E. W. Cliver, D. F. Webb, R. A. Howard: Solar Phys. **187**, 89 (1999)
7. E. W. Cliver, S. W. Kahler, D. V. Reames: Astrophys. J. **605**, 902 (2004a)
8. E. W. Cliver, N. Nitta, B. J. Thompson, J. Zhang: Solar Phys. **225**, 105 (2004b)
9. I. J. D. Craig, Y. E. Litvinenko: Astrophys. J. **570**, 387 (2002)
10. N. Gopalswamy: Planetary and Space Sci. **52**, 1399 (2004a)
11. N. Gopalswamy: A Global Picture of CMEs in the Inner Heliosphere, In: *The Sun and the Heliosphere as an Integrated System*, ed by G. Poletto and S. Suess (Kluwer, Berlin Heidelberg New York 2004b), pp. 201–252
12. N. Gopalswamy: Interplanetary Radiobursts. In *Solar and Space Weather Radiophysics*, ed by D. Gary and C. Keller (Kluwer, Berlin Heidelberg New York 2004c), pp. 305–333
13. N. Gopalswamy, M. L. Kaiser, R. P. Lepping, et al.: J. Geophys. Res. **103**, 307 (1998)
14. N. Gopalswamy, A. Lara, M. L. Kaiser, J.-L. Bougeret: J. Geophys. Res. **106**, 25261 (2001a)
15. N. Gopalswamy, S. Yashiro, M. L. Kaiser et al.: J. Geophys. Res. **106**, 29219 (2001b)
16. N. Gopalswamy, S. Yashiro, M. L. Kaiser, et al.: Astrophys. J. Lett. **548**, L91 (2001c)
17. N. Gopalswamy, S. Yashiro, M. L. Kaiser et al.: Geophys. Res. Lett. **29**, 106 (2002a)
18. N. Gopalswamy, S. Yashiro, G. Michalek et al.: Astrophys. J. **572**, L103 (2002b)
19. N. Gopalswamy: Geophys. Res. Lett. **30**, SEP1-1 (2003)
20. N. Gopalswamy, S. Yashiro, A. Lara et al.: Geophys. Res. Lett. **30**, SEP 3–1 (2003a)
21. N. Gopalswamy, S. Yashiro, M. L. Kaiser, R. A. Howard: Adv. Space Res. **32 (12)**, 2613 (2003b)
22. N. Gopalswamy, S. Nunes, S. Yashiro, R. A. Howard: Adv. Space Res. **34 (2)**, 391 (2004a)
23. N. Gopalswamy, S. Yashiro, S. Krucker et al.: J. Geophys. Res. **109**, A12105 (2004b)
24. N. Gopalswamy, S. Yashiro, S. Krucker, R. A. Howard: CME Interaction and the Intensity of Solar Energetic Particle Events. In: *Coronal and Stellar Mass ejections, IAU Symposium No. 226*, ed by K. Dere, J. Wang and Y. Yan, pp. 367–373 (2005a)
25. N. Gopalswamy, E. Aguilar-Rodriguez, S. Yashiro et al.: J. Geophys. Res. **110**, A12, A12S07 (2005b)
26. J. T. Gosling, E. Hildner, R. M. MacQueen, et al.: Solar Phys. **48**, 389 (1976)
27. D. Haggerty, E. Roelof: Astrophys. J. **579**, 841 (2002)
28. S. Hoang, J.-L. Steinberg, G. Epstein, et al.: J. Geophys. Res. **85**, 3419 (1980)
29. A. J. Hundhausen: J. Geophys. Res. **98**, 13177 (1993)
30. S. W. Kahler: Astrophys. J. **261**, 710 (1982)
31. S. W. Kahler: J. Geophys. Res. **106**, 20947 (2001)
32. S. W. Kahler, E. Hildner, M. A. I. van Hollebeke: Solar Phys. **57**, 429 (1978)
33. S. W. Kahler, N. R. Sheeley, Jr., R. A. Howard, et al.: Solar Phys. **93**, 133 (1984)

34. S. W. Kahler, N. R. Sheeley, Jr., R. A. Howard, et al.: J. Geophys. Res. **89**, 9683 (1984)

35. M.-B. Kallenrode: J. Phys. G: Nucl. Part. Phys. **29**, 965 (2003)

36. K.-L. Klein, G. Trottet, H. Aurass, A. Magun, Y. Michou: Adv. Space Res. **17** (**4–5**), 247 (1996)

37. K.-L. Klein, J. I. Khan, N. Vilmer et al.: Astron. Astrophys. **346**, L53 (1999)

38. S. A. Knock, I. H. Cairns, P. A. Robinson, Z. Kuncic: J. Geophys. Res. **108** (**A3**), SSH6-1 (2003)

39. M. Krogulec, Z. E. Musielak, S. T. Suess, S. F. Nerney, R. L. Moore: J. Geophys. Res. **99**, 23489 (1994)

40. A. Lara, N. Gopalswamy, S. Nunes, G. Muñoz, S. Yashiro: Geophys. Res. Lett. **30**, 8016 (2003)

41. R. P. Lin: Solar Flare Particles, in: *From the Sun: Auroras Magnetic Storms, Solar Flares, Cosmic Rays*, ed by S. T. Suess and B. T. Tsurutani, (American Geophysical Union, Washington, DC. 1998), pp. 91–102

42. R. P. Lin, S. Krucker, G. J. Hurford et al.: Astrophys. J. Lett. **595**, L69 (2003)

43. J. L. Lovell, M. L. Duldig, J. E. Humble: J. Geophys. Res. **103**, 23733 (1998)

44. H. H. Malitson, J. Fainberg, R. G. Stone: Astrophys. Lett. **14**, 111 (1973)

45. G. Mann, A. Klassen, C. Estel, B. J. Thompson: In: *Proc. of 8th SOHO Workshop*, ed by J.-C. Vial and B. Kaldeich-Schürmann. (ESA, 1999), p. 477

46. G. Mann, A. Klassen, H. Aurass, H.-T. Classen: Astron. Astrophys. **400**, 329 (2003)

47. G. M. Mason, J. E. Mazur, J. R. Dwyer: Astrophys. J. **525**, L123 (1999)

48. G. J. Nelson, D. B. Melrose: Type II bursts. In: *Solar Radiophysics*, ed by D. J. McLean and N.R. Labrum, (Cambridge University Press, Cambridge 1985), pp. 333–359

49. G. Newkirk: Astrophys. J. **133**, 983 (1961)

50. R. Payne-Scott, D. E. Yabsley, J. G. Bolton: Nature **160**, 256 (1947)

51. U. R. Rao, K. McCracken, R. P. Bukata: J. Geophys. Res. **72**, 4325 (1967)

52. D. V. Reames: Space Sci. Rev. **90**, 413 (1999)

53. D. V. Reames:Astrophys. J. **571**, L63 (2002)

54. M. J. Reiner, M. Karlicky, K. Jiricka et al.: Astrophys. J. **530**, 1049 (2000)

55. N. R. Sheeley, Jr., R. A. Howard, D. J. Michels et al.: Astrophys. J. **279**, 839 (1984)

56. G. M. Simnett, E. C. Roelof, D. K. Haggerty: Astrophys. J. **579**, 854 (2002)

57. E. J. Smith, R. G. Marsden, A. Balogh et al.: Science **302**, 1165 (2003)

58. R. T. Stewart, R. A. Howard, F. Hansen et al.: Solar Phys. **36**, 219 (1974)

59. Z. Svestka, L. Fritzova-Svestkova: Solar Phys. **36**, 219 (1974)

60. J. Torsti, L. Kocharov, J. Laivola: Astrophys. J. **573**, L59 (2002)

61. B. Vrsnak, V. Ruzdjak, P. Zlobec, H. Aurass: Solar Phys. **158**, 331 (1995)

62. B. Vrsnak, A. Warmuth, M. Temmer et al.: Astron. Astrophys. **448**, 739 (2006)

63. J. P. Wild, L. L. McCready: Austral. J. Sci. Res. **A3**, 387 (1950)

64. R. Woo, J. W. Armstrong: Nature **292**, 608 (1981)

Particle Acceleration at the Earth's Bow Shock

David Burgess

Astronomy Unit, Queen Mary, University of London
D.Burgess@qmul.ac.uk

Abstract. Shocks can be viewed as sites where upstream bulk flow energy is converted into downstream thermal energy. Collisionless shocks have the further property that they are sites of particle acceleration, as some fraction of the flow energy is diverted to a small number of energetic particles. Shocks are thus important in many astrophysical and solar environments. The best studied example is the Earth's bow shock, which has the benefit of high time resolution, in situ multipoint measurements. Models and mechanisms developed for the Earth's bow shock are useful for understanding the behaviour of shocks in the corona and solar wind. The Earth's bow shock is curved, and the main controlling factor of its structure is the angle between the magnetic field and shock normal, which is used to classify the shock as quasi-parallel or quasi-perpendicular. We review observations of accelerated electrons at the quasi-perpendicular shock and the standard model of acceleration, namely adiabatic reflection. We describe the weaknesses of this model and suggest a new mechanism, based on simulations, which predicts power law energy spectra. The quasi-parallel shock is a region of large amplitude turbulence, containing coherent structures and wavetrains. Cluster multi-point observations of these large amplitude pulsations are presented as a review of the present understanding of this type of shock. In terms of particle acceleration, quasi-parallel shocks are important since they are believed to be the best examples of sites for Fermi acceleration. Cluster observations are reviewed indicating unambiguously that energetic particle diffusion, a fundamental assumption of the Fermi process, is operating at the quasi-parallel bow shock. Finally, a class of transient events observed at the bow shock, Hot Flow Anomalies, is described. The causative mechanism and implications for particle acceleration are discussed, and a possible role in other heliospheric environments is suggested.

1 Introduction

Collisionless shocks are a key component in astrophysical systems, since they can transfer bulk flow kinetic energy to a small population of highly energetic particles. For radio emission mechanisms this is vital since they almost always depend on the presence of some kind of energized electron population. Of course, we also know from direct observations of the solar wind that shocks do exist, are associated with energetic particles, and are responsible for some

D. Burgess: *Particle Acceleration at the Earth's Bow Shock*, Lect. Notes Phys. **725**, 161–190 (2007)
DOI 10.1007/978-3-540-71570-2_8

kinds of radio emission (e.g. coronal and interplanetary Type II bursts). We have high resolution observations of interplanetary shocks from the inner to outer heliosphere, and highly detailed observations of the Earth's bow shock over a wide range of solar wind conditions. Our understanding of the physics of collisionless shocks has been advanced greatly by analysis of bow shock observations. In terms of observations, the Earth's bow shock is unique. In many ways it is used as a template for understanding shocks in other astrophysical phenomena, from the solar wind termination shock to supernova remnant shocks.

Shocks are often invoked to explain various coronal solar radio phenomena, where in situ measurements are not possible. In such cases, it is important to understand the underlying processes of shock acceleration, and the nature of processes at the Earth's bow shock (where detailed in situ measurements are possible) so that one can apply results appropriately. For example, coronal plasma conditions can be very different from those at the bow shock, so the regime of applicability needs to be carefully considered.

We have a unique view of the Earth's bow shock in terms of time resolution, range of parameters and availability of multi-point data. Studies have demonstrated how important the upstream conditions are for determining the observed shock type, and at the bow shock we have data available from upstream solar wind monitors. Bow shock measurements have progressed with data from the Cluster mission, with four, fully instrumented spacecraft. Multi-point measurements are necessary to disentangle the space-time ambiguity when a change is observed at only one point.

A shock forms upstream of an obstacle in a supersonic flow. The Earth's bow shock stands upstream of the magnetopause, so that the bow shock surface is approximately cylindrically symmetric about the Sun-Earth line, with a paraboloid form on the sunward side. The shock's position is between 10–20 R_E from the Earth, and the stand-off distance from the magnetopause is several R_E. The actual bow shock position changes rapidly in response to changes in the solar wind. Most observations of the bow shock are due to motion of the shock over the spacecraft.

The bow shock is typically a high Mach number shock with an Alfvén Mach number M_A of order 5–10, and similar for the fast magnetosonic Mach number. The Mach number falls off towards the flanks, since only the normal component of velocity plays a role in the shock jump conditions. At any instant the interplanetary magnetic field (IMF) direction is roughly constant over the surface of the shock. (Although it can change abruptly as the solar wind magnetic field convects over the bow shock at the solar wind speed.) The mean IMF direction is given by the Parker spiral angle. Since the shock surface is curved, this means that the angle between the shock normal and the upstream magnetic field direction, θ_{Bn}, varies over the surface. Shocks with θ_{Bn} in the ranges $0 \lesssim \theta_{Bn} \lesssim 45$ and $45 \lesssim \theta_{Bn} \lesssim 90$ are known as quasi-parallel and quasi-perpendicular, respectively. The angle θ_{Bn} is the major controlling factor for the structure of the shock and the thermalization and acceleration processes operating there [1].

Quasi-perpendicular shocks have a laminar appearance in flow and magnetic field profiles, with a foot-ramp-overshoot structure, seen in the field (passing from upstream to downstream). The structure and ion thermalization is controlled by the ion dynamics in this field structure. An important fraction (10–20%) of ions are reflected at the ramp, where the gradient is largest. The magnetic geometry means they gyrate around the field immediately upstream, where they form a "foot" structure in density, flow and field which extends in front of the shock. On their re-encounter with the shock, the reflected gyrating ions pass downstream, where they contribute to the overshoot structure where the field is larger than its eventual downstream value. The reflected ions' large gyrational velocity, after scattering at or downstream of the shock, results in the thermalization required at the high Mach number shock (see, e.g., [1]). A review of the quasi-perpendicular shock based on Cluster observations is given in [2].

The quasi-parallel shock has a radically different, far more turbulent structure and different processes operate there. In particular, the magnetic geometry allows particle motion upstream from the shock for particles which have gained sufficient energy by either reflection or escape from the downstream heated population. The structure that results is closely linked to this property and the related presence of energetic ions at the shock; this is discussed in greater detail in Sect. 3.

At the Earth's bow shock the solar wind is collisionless, so that any particle with appropriate energy and pitch angle can escape the shock and travel upstream. The foreshock is the region ahead of the shock filled with energetic particles accelerated at the shock, and waves driven by their unstable particle distributions. For any given shock, the properties of the particles that can escape depend on θ_{Bn}, since the minimum velocity for escape from the shock corresponds to the speed of the shock along the upstream field lines, which varies as $\sec\theta_{Bn}$. In addition, the particles escaping upstream also suffer the cross-field drift imposed by the convection electric field of the solar wind. This acts to convect the escaping particles in the solar wind direction. If particles with a range of parallel velocities leave from a location on the shock, their trajectories will be dispersed by a "velocity filter" effect, since those with higher parallel velocity travel more closely aligned with the magnetic field. The foreshock at the Earth's bow shock is also affected by the curvature of the shock surface. Different points within the foreshock are connected magnetically to different locations on the shock surface, which have corresponding different values of θ_{Bn}.

The energization of different species depends on the θ_{Bn} and Mach number of the shock, so that accelerated electron and ion sources are distributed differently over the bow shock surface. This effect, together with that of velocity dispersion, leads to distinct spatial regions within the foreshock dominated by different species and energies. Thus commonly the foreshock is typically divided into an "electron foreshock" and an "ion foreshock," with distinct characteristics in terms of energetic particles and associated waves. These various effects, as if not enough, are further complicated by the feedback of waves, generated by the accelerated particles, on the whole shock system. For example, ion and electron foreshocks have very different characteristics of low frequency magnetic waves, and this plays a role in the actual structure of the bow shock.

The roles of the velocity filter effect and the shock escape criterion have been discussed by, e.g., [3, 4, 5]. The range of waves found in the terrestrial foreshock has been reviewed in [6]. When radio emission at shocks is discussed it usually refers to emission in the foreshock via wave processes driven by electrons accelerated at the shock. So, understanding the shock, its structure and how it accelerates particles is usually just one step towards a theory for radio emission at the shock.

Although we have been enthusiastic about the advantages of the bow shock for studying shock processes, it is also fair to point out some disadvantages. The bow shock parameters (Mach number, plasma beta etc.) are limited to those found at 1AU, and, although this covers a large range, it means that direct comparison with all coronal conditions is not possible, and, indeed, care is needed to extrapolate from 1AU to further out in the heliosphere. The curvature of the bow shock surface, together with the fact that it is relatively small means that θ_{Bn} and Mach number vary relatively fast (compared with some plasma scales) so that there can be interaction between neighbouring parts of the shock with very different parameters. Usually this interaction is mediated by the foreshock. This has to be borne in mind when interpreting foreshock phenomena or applying them to model other situations. For example, interplanetary shocks, which, because of their expansion factor, can be considered as predominantly planar, have a completely different configuration.

This paper reviews some open issues from studies of particle acceleration at the bow shock, and presents a selection of recent contributions. Some topics are relevant to radio physics, others important if only to rule out certain possibilities. Other topics touch on fundamental questions of shock physics which are vital to understanding energy flow in astrophysical systems, whether in the corona, interplanetary space or beyond.

2 Electron Acceleration at the Quasi-perpendicular Bow Shock

2.1 Motivations

In the case of electrons there are several obvious reasons why the shock acceleration process must be studied and understood. Foremost, and the reason why the bow shock is appropriate to pursue such studies, is the direct observation of energetic electrons upstream of collisionless shocks. The region containing non-thermal, energized electrons is called the electron foreshock. Since the speed of energetic electrons is much greater than the solar wind convection speed, the upstream edge of the electron foreshock is usually taken as the magnetic field tangent surface to the bow shock (i.e., the locus of points on magnetic field lines which are tangent to the bow shock surface). Direct observations cover the energy range from suprathermal (above several thermal speeds) to tens of keV, and are discussed in the next section.

Apart from direct observations of energetic electrons, the foreshock is also a known radio source at twice the local plasma frequency (the so-called electromagnetic $2f_p$ radiation). In addition, the electron foreshock contains Langmuir turbulence and other plasma emissions which indicate the presence of electron beams. This indirect evidence of electron acceleration is a further motivation to study the processes responsible, since the radio emission will have properties controlled, to a greater or less degree, by the exciting population of electrons. This can be taken in two directions: either to infer electron properties from the radio and plasma emission, or to use the data from the bow shock to model distant phenomena (solar, heliospheric, astrophysical, etc.) and thus to understand their radio emissions (e.g., [7, 8, 9]).

Finally, although the electron energies at the bow shock are not particularly high, the detail of in-situ measurements allows us to gain insight into the processes whereby particles are extracted from the thermal population, raised in energy at the shock, and made available for "injection" in some other, larger scale and higher energy, acceleration process. The hope here is that a convincing scenario can be developed for a general acceleration process that starts at thermal energies and which can be traced up in energy to that of the most energetic cosmic rays. Of course, whether or not shock acceleration can be invoked successfully at every astrophysical example of particle acceleration is another matter. But the bow shock gives us a near-ideal environment for determining the strengths and limitations of shock acceleration.

2.2 Observations at the Bow Shock

Observations key to our understanding of the electron foreshock were published by Anderson et al. [10] using ISEE data. Figure 1, shows time series of fluxes of electrons in the energy bands around >16 keV, 5.3 keV, and 1.5 keV. The period shown can be interpreted as a passage through the foreshock, starting in the magnetically unconnected solar wind. Note, that in this case the foreshock moves over the spacecraft due to a change in the solar wind magnetic field direction. The dashed line indicates the time at which the spacecraft is estimated to have passed through the upstream tangent surface, based on the onset times for both energetic electrons and protons. There is clearly velocity dispersion, in that the > 16 keV electron flux increases before that of the lower energy channels. This can be explained by simply assuming a source of energetic electrons at the bow shock. All particles, irrespective of

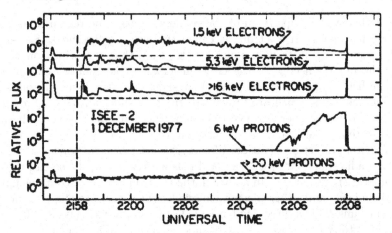

Fig. 1. ISEE-2 observations of upstream electrons. The upstream edge of the foreshock is marked with a *dashed line*. (From [10])

energy, suffer the same cross-field $E \times B$ drift (from the convection electric field), but particles with higher energy have higher parallel velocity, and so the angle between the velocity (in the bow shock frame) and the magnetic field is smaller than for particles of lower energy. The effect is that particles of lower energy are more readily swept downstream. The region just behind the tangent field line is filled with particles of higher energy, but not the lower energy particles, which are convected away from the tangent field line.

This interpretation (as given in [10]), has been refined and elaborated by many authors, e.g., [3, 4, 5]. Incidentally, the spike of energetic electrons at 2208 UT (Fig. 1) is associated with a rapid rotation of the magnetic field direction causing the foreshock to move away. The increase in the 6 keV proton flux at 2205 is probably associated with field aligned ion beams [10].

However, Fig. 1 reveals another interesting thing: The fluxes at >16 keV and 5.3 keV actually decrease as the 1.5 keV fluxes rise. Thereafter the 1.5 keV fluxes remain present, indicating that they are present throughout the foreshock. Anderson et al. interpreted this as indicating that the energetic electrons were only accelerated near where the tangent field lines touched the shock surface, i.e., where the shock normal angle is in the range 1°–5° from perpendicular. They estimated that this corresponded to a region of width $\sim 10^4$ km on the shock surface. As the accelerated electrons travel away from the bow shock they fill a thin sheet just behind the tangent surface. The temporal changes in fluxes are thus primarily a result of changes in the interplanetary magnetic field (IMF) direction which causes the electron foreshock to sweep back and forth across the spacecraft. The spacecraft motion relative to the bow shock is relatively unimportant on the time-scales of tens of minutes.

The rapid variation of electron fluxes at the highest energies has meant that the observations of Anderson et al. [10] have not been much improved since. Results from an experiment with similar high time resolution are presented in [11]. The bulk of electron observations in the foreshock have been made with instruments designed for thermal and suprathermal energies. This has meant that the details of the distribution functions are known in detail at lower energies and at low time resolutions. Feldman et al. [12] present observations showing velocity dispersion at lower energies up to ~1 keV. Of particular relevance to the question of plasma wave instability is the issue whether a secondary peak in the reduced distribution can be observed. Such a peak (bump-on-tail) is predicted by models for the observed plasma emission, e.g., [3], and inferred from the observations [10].

Fitzenreiter et al. [13] first presented evidence for bump-on-tail distributions, however these are only sometimes resolved. Further work [5] presented pitch angle distributions of foreshock electrons. In particular, it was seen that, when deep in the foreshock, there was evidence for loss-cone depletion at low (i.e., thermal) energies. Pitch angle variations, and the similarity of incoming and backstreaming fluxes are an indication that magnetic mirroring may play an important role in the electron reflection process, as suggested by the authors of these observational papers. The link between reflection and energization is dealt with in the next section.

An alternative observational viewpoint was put forward by Gosling et al. [14] who carried out a survey of bow shocks seen by the ISEE spacecraft, and made the point that suprathermal electrons with energies up to about 20 keV are commonly measured downstream of quasi-perpendicular shocks, but rarely behind the quasi-parallel bow shock. Furthermore, the suprathermal electron flux appears as a high energy tail with a power law form (with exponent in the range 3–4), which smoothly emerges from the thermal distribution. The fluxes peak just downstream of the shock ramp and the highest energies (>5 keV) appear in the region of the shock overshoot. A field-aligned backstreaming component is observed in the ramp and this was interpreted as escaping electrons which have been energized at, and downstream of the shock. Gosling et al. argued that simple magnetic mirroring could not explain the observed downstream fluxes of suprathermal electrons. They argued that leakage from downstream would explain their observations of backstreaming electrons in the foreshock, but they did not specify a mechanism to produce a suitably energized suprathermal distribution.

2.3 Mechanisms

Given the magnetic field jump at the quasi-perpendicular shock, and the observational evidence for reflection, it seems natural to examine the possibility that magnetic mirroring is responsible. The first analytical studies were presented by Leroy and Mangeney [15] and Wu [16]. Both sets of authors calculated the distribution function and moments (e.g., energy and flux) of electrons reflected at a quasi-perpendicular shock. The steps in the model are as follows: Only time-steady fields and a planar shock are considered. In the shock normal incidence frame, NIF, (in which the shock is at rest and the upstream flow is directed along the shock normal into the shock) the interaction between particles and fields is complicated by the convection electric field. Thus the first step is to transform into the so-called de Hoffman-Teller frame (HTF), in which the shock is at rest, and the upstream flow is along the magnetic field direction. In the HTF there is zero convection electric field, so that, provided the fields remain time-stationary, particles will conserve their energy [1, 17]. This radically simplifies the description of the particle-shock interaction.

If the gradients in the shock are sufficiently small, and, again we emphasize, if the particle motion is scatter-free, the motion is adiabatic, so that the magnetic moment $\mu = \frac{m(v^{HT})^2}{2B}$ is constant, where the superscript HT indicates the frame in which the quantity is evaluated. For a given maximum of value of B, those particles from the incident distribution with sufficiently large pitch angle will be reflected, i.e., have their component of $v_\parallel^{HT}$ reversed. By Liouville's theorem the phase space density at the corresponding point in velocity space will be the same as the incident distribution at the point whence the reflected particle originated: $f_R(-v_\parallel^{HT}, v_\perp^{HT}) = f_0(v_\parallel^{HT}, v_\perp^{HT})$. This is illustrated in Fig. 2. There is a complication evident in the figure.

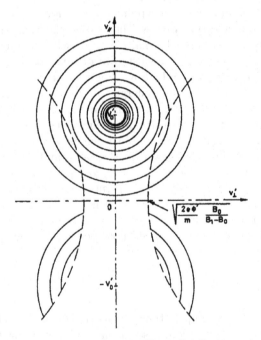

Fig. 2. Velocity space contours of an incident distribution at a shock and the portion reflected due to magnetic mirroring, modified by the effect of the cross-shock potential. (From [15])

Quasi-perpendicular shocks also have a normal directed electric field which is not removed by the transformation to the HTF. This electric field is required at the quasi-perpendicular shock to reflect some of the thermal ions and to control the electron heating. Accordingly the energy equation for an electron is

$$\frac{1}{2}m[v_\parallel^{HT}(x_i)]^2 = \frac{1}{2}m[v_\parallel^{HT}(x)]^2 + \Psi(x)$$

where the shock normal is along x, and x_i is an initial position upstream of the shock. $\Psi(x)$ is a pseudo-potential function $\Psi(x) = \mu[B(x) - B(x_i)] - e\Phi^{HT}(x)$, where $B(x)$ and $\Phi(x)$ are the magnetic field magnitude and electrostatic potential in the HTF as functions of position x. Recall that the shock is assumed one-dimensional so the shock fields can only vary with x, and that the magnetic field can be assumed unchanged by the transformation to the HTF due to the nature of its Lorentz transformation.

The sense of the electrostatic potential is to reflect upstream ions, and contain downstream electrons. Its effect on incident electrons is to make lower energy particles (in the HTF), that would otherwise have been reflected, pass downstream. The final result is that the reflected distribution has the form of a ring beam, with sharp edges on the loss-cone edges.

In the reflection process the particle energy is conserved in the HTF, but the parallel component of velocity is reversed. It is this change of velocity, as seen in the NIF (or, e.g., the upstream plasma frame) which leads to the energization of the particles. The de Hoffman-Teller transformation velocity can be found from purely geometrical considerations. Figure 3 shows velocity space in both NIF and HTF. The de Hoffman-Teller transformation velocity V_{HT} is the velocity along the shock surface which makes the incident flow field-aligned in the HTF, and also keeps the shock at rest. From the basic

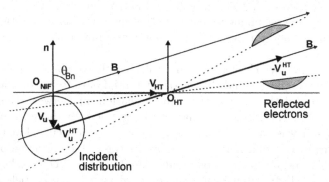

Fig. 3. Relationship between NIF and HTF velocity spaces. V_{HT} is the HTF transformation velocity. The origins of the frames are marked O_{NIF} and O_{HT}, respectively. One contour from the incident distribution is shown, together with the corresponding reflected portion satisfying a magnetic mirroring criterion (*dashed line*). Note that the scale is exaggerated so that the transformation velocity can be seen clearly, and so that the effect of the cross shock potential is not evident

geometry $V_{HT} = V_u \tan \theta_{Bn}$, where V_u is the upstream flow speed in the NIF. It is clear that V_{HT} increases dramatically as θ_{Bn} approaches 90°. Technically the HTF cannot be found for an exactly perpendicular shock, but the frame transformation can be used for angles very close to 90°.

In Fig. 3 a contour of an incident distribution function is also shown and the part of the distribution with the correct pitch angles to satisfy a magnetic mirroring criterion, which will depend on the maximum magnetic field seen by the electrons. The reflected portion is also shown. (Note, that for clarity, this contour is shown at an unrealistically small peculiar velocity; at the bow shock the electron thermal speed is much greater than the upstream flow speed.) The average energy, in the NIF, of the reflected portion is approximately the energy associated with twice the incident flow speed in the HTF, i.e., $\sim \frac{4}{\cos^2 \theta_{Bn}} \left(\frac{1}{2} m V_u^2 \right)$.

As θ_{Bn} approaches 90°, V_{HT} increases, and so too does the energy of the reflected particles. However, as can be seen from Fig. 3, as V_{HT} increases, if the pitch angle for mirroring remains constant, the portion that will mirror increasingly comes from the outer parts of the incident distribution (i.e., the parts with higher energy in the flow frame). Thus the reflected density will depend strongly on how the incident distribution falls off above thermal energies. So, we are left with a paradox: Acceleration by reflection due to magnetic mirroring will give the highest energies for shocks closest to perpendicular, but the same shocks give the lowest reflected fraction.

This was realized in the first theories [15, 16] which calculated reflected density fraction and energy analytically for a Maxwellian distribution and a two temperature (core plus halo) distribution, as often found in the solar wind. Later modelling work used a Lorentzian distribution which has the advantage of higher phase space density at high energies compared to a Maxwellian. Models have also been developed for the variation of the reflected electron distribution within the foreshock by using Liouville mapping [5]. Cairns et al. [18] used the same technique to calculate the electron beam properties, Langmuir growth rate and the available free energy as a function of position within the foreshock.

In determining which electrons are reflected the combined effects of the magnetic field amplitude and the electric potential have to be considered. So, in order to model the process the spatial relationship between the electric potential and the magnetic field has to be assumed. In the first modelling work [15, 16] it was assumed that the maxima of both occurred at the same position within the shock. In order to avoid this arbitrary assumption 1D hybrid simulations of the quasi-perpendicular shock have been used to supply (time-varying) electric and magnetic fields, in which test particle electron trajectories were followed [19].

A hybrid simulation models the plasma as a set of macro-particles for the ion (i.e., proton) component, and a massless, charge-neutralizing fluid for the electrons. This type of simulation has had great success modelling the overall structure of quasi-perpendicular shocks and the ions dynamics and distributions found at such shocks. The hybrid simulations are self-consistent in so far as the fields are generated by the plasma, as it is modelled. Energetic electron test particles are followed in these fields, with the assumption that the overall shock fields would not be qualitatively changed by this population if its contribution to charges and currents were included.

The work described in [19] also dropped the assumption of adiabatic motion, but nevertheless it essentially confirmed the early analytic adiabatic studies, and also their assumption of the colocation of magnetic field and electrostatic potential maxima. A further result was that the distance travelled by electrons in the reflection process could be unexpectedly large. This motion consists of both drift along the motional electric field direction (which leads to energy gain in the NIF) and motion parallel to the magnetic field, which can be large due to the high velocity of the electrons (and, of course, the assumption of scatter-free propagation). It should be noted that the reflection time can be relatively long, since for particles reflected near the field maximum the time is associated with the transit time of a field line through the shock structure, which is tied to the bulk motion, and which at the bow shock is considerably smaller than, e.g., the electron thermal speed.

Since the Earth's bow shock has a relatively small radius of curvature, this latter result prompted an investigation of the effects of curvature on the simple scatter-free, magnetic mirroring scenario [20]. Again, the electric and magnetic fields were supplied by 1D hybrid simulations, but modified in such a way as to model a synthetic curved shock, over a limited range of θ_{Bn} 86°–90°. An incident electron population with a Lorentz distribution was used, and it was found that, most importantly, the curvature did not completely invalidate the model of energization by magnetic mirroring reflection. Indeed, the curvature plays a role in producing a flux focussing effect so that electrons entering at close to $\theta_{Bn} = 90°$ drifted and exited at slightly lower values of θ_{Bn}. This produces a flux maximum at an exit position of $\theta_{Bn} \sim 87°$. Although the curved shock model could reproduce the order of magnitude of the observed fluxes, there were, however, discrepancies with the form of the energy spectra. A further issue which was highlighted was the dependence of the results on the choice of incident distribution function.

The effect of a curved shock, with finite width, was also investigated by Vandas [21] using analytic solutions for test particle electrons in suitably chosen shock profiles. This work was extended [11] to include comparisons between modelled and observed upstream anisotropies and energy spectra. It was found that the effect of curvature was to shift the maximum in the reflected flux away from exactly perpendicular, by a few degrees; this confirmed the earlier results using hybrid simulations [20]. Vandas [11] used test particle calculations to model what a spacecraft would observe as it approached and crossed the bow shock, and also made comparisons with observations from the Prognoz 10 spacecraft. The overall conclusion was that the similarities between observed and modelled properties were strong, but essentially qualitative, i.e., in terms of order of magnitude of fluxes and energies. The discrepancies were that the modelling predicted higher anisotropies, both upstream and downstream, than observed. And, crucially, the observed power law index (e.g., as reported for downstream [14]) did not agree with the modelling results. It was concluded that the isotropic velocity distribution observed downstream indicated another process, such as pitch angle scattering, was operating in addition to shock drift acceleration via reflection.

We are concentrating intentionally on the acceleration processes responsible for the most energetic electrons seen at the Earth's bow shock, since they are associated with the $2f_{pe}$ radio emission region. However, the electron foreshock also includes lower energy backstreaming components at progressively "deeper" points, i.e., locations magnetically connected to lower θ_{Bn}. These components could be simply heat flux escaping from the downstream heated distribution, and/or a reflected population formed by magnetic mirroring (as used to explain the energetic electrons at the foreshock edge). Indeed observations of foreshock electrons at moderate energies often show features around the magnetic loss-cone and these have been modelled by using Liouville mapping (i.e., conservation of phase space density along scatter-free particle trajectories) into the foreshock [5]. Another example of the use of electron acceleration by adiabatic reflection is that of Knock et al. [7] who modelled the foreshock of a curved shock in order to parameterize electron beam properties, which were then used in a model of Type II solar radio emission.

One of the key assumptions of electron acceleration by reflection is that the particle motion is scatter-free, so the magnetic moment is conserved. On the other hand ideas of non-stationarity of the quasi-perpendicular shock, and consequent non-conservation of magnetic moment, have been explored theoretically by a number of authors, such as in [22, 23]. There is also a body of simulation work that indicates that the quasi-perpendicular shock is not time-stationary, but exhibits an over-turning solution (e.g., [24]). These simulations are full-particle, with kinetic macroparticles for both ions and electrons, so that in principle the interaction between the shock fields and the electron dynamics will be fully, and properly, captured. It should be noted, that full particle simulations, due to computational constraints, have to use non-physical parameters, such as, for example, an unrealistically small proton to electron mass ratio. Because of these limitations, there is some debate as how to successfully apply the results from such simulations [25].

A full-particle simulation study showing the formation of bursts of electrons is presented in Lembège and Savoini [26]. Not only does the shock front non-stationarity (or reformation of the shock ramp) produce bursts of reflected particles, but also variation of the shock surface in the transverse direction produces spatially localized clumps of reflected electrons. These simulations present a picture of electron energization very much in contrast to the simple scenario of reflection by magnetic mirroring (although conservation of magnetic moment might be important for the electrons seen reflected in the simulations). We note that these simulations are for a shock with $\theta_{Bn} = 55°$, and so not directly applicable to explaining the observations of the most energetic electrons seen at the foreshock edge. Further simulations [27] have also been carried out of a curved, two-dimensional shock, which modelled the formation of the entire electron foreshock. A number of qualitative features of the observations were found, such as a loss-cone signature in the reflected electron distributions. However, limited particle statistics restricted any conclusions about the electrons seen at the foreshock edge.

2.4 Shock Structure and Particle Scattering

We have seen how factors such as curvature and non-stationarity have been shown to be possible factors which affect the simple model of adiabatic reflection. We also note the comment in [11] that an additional process, such as pitch angle scattering, must be operating in order to explain the observed downstream suprathermal electron isotropy and power law energy spectra. Krauss-Varban [28] extended the quasi-2D study of the interaction of electrons at a curved shock by examining the effect of adding ad hoc pitch angle scattering. In particular, in this study it was found that the process could be very efficient in redistributing energized electrons, so that significant fluxes could be observed over a broader range of θ_{Bn} than predicted by the simple adiabatic theory. However, the conclusions were limited by the ad hoc nature of the assumed pitch angle scattering.

Although high frequency waves associated with electron dynamics (such as lower hybrid or electron whistler waves) could be, and probably are a source of pitch angle scattering, a novel mechanism has recently been suggested [29]. It was noted that in two-dimensional hybrid simulations the shock front exhibits transverse structure, consisting of ripples, seen most clearly in the shock normal component of the magnetic field. This particular structure is only seen in the simulations when the simulation plane contains the upstream magnetic field, so that fluctuations propagating along the field direction can develop and propagate. Lowe and Burgess [29] pointed out that the motion along the field lines could mean that some electrons are trapped within the magnetic structures as they pass downstream. It was suggested that, in principle, this could lead to effective Fermi acceleration. Indeed, energy spectra were shown for test particle electrons (moving in the fields from the 2D hybrid simulations) demonstrating that downstream an approximately power law energy dependence was found. This was suggested as a possible explanation for the observations [14].

In these two-dimensional simulations the power law energy spectra were only seen for cases with the upstream magnetic field in the simulation plane. If the simulation was arranged so that the upstream field was perpendicular to the simulation plane, then the shock still formed, but it lacked the field structuring, and much more resembled the results of one-dimensional hybrid simulations. This indicates that structuring along the field direction (or transverse to the shock normal) plays a crucial role in the electron dynamics.

Further details of the structuring of the shock were presented in [30] which investigated the properties of the shock surface ripples seen in the two-dimensional hybrid simulations. Figure 4 shows the shock normal component B_x for a shock with $\theta_{Bn} = 88°$ as a two-dimensional intensity map. Also shown is the profile of B averaged over the transverse direction; this is essentially the same as seen in one-dimensional simulations, which indicates that the overall dynamics of the shock are not radically different in two-dimensions. An interesting result was that the main component of the ripples propagated over the shock surface with a range of wave number k, but with a constant phase speed, which was found to be approximately the Alfvén speed at the position of the shock overshoot. This is shown in Fig. 5 which shows the $\omega - k$ power distribution in B_x field slices at the overshoot. Note the normalization of k uses the upstream Alfvén speed. The peak power follows a straight line in the $\omega - k$ plane, indicating constant phase speed. This property, together with the near exponential fall off of the ripple amplitude away from the overshoot position, gives a strong indication that the ripples are a form of surface wave. There is another, weaker, component which appears to have a wave vector oblique to the magnetic field and shock normal; this component is more apparent in B_z, the out of coplanarity plane component, and likely to be related to whistler wave packets in the foot and ramp region.

The rippling of the shock, i.e., nonstationary structuring along the magnetic field lines, has a major impact on the dynamics of suprathermal electrons

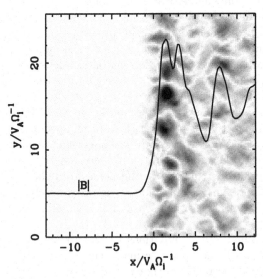

Fig. 4. Intensity map of shock normal component of magnetic field from a two dimensional hybrid simulation of a shock with $\theta_{Bn} = 88°$, $M_A = 6$. The magnetic field profile averaged over the y direction is also shown. (From [30])

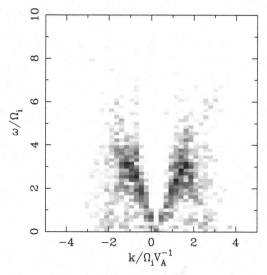

Fig. 5. The $\omega - k$ distribution of power for the B_x component recorded at the position of the shock overshoot. Only propagation transverse to the shock normal is considered. The *grey scale* covers a factor of 1000 logarithmically. (From [30])

at the shock, as shown in [29]. Figure 6 shows the upstream and downstream energy distributions resulting from an injected population of test particle electrons with initial energy 100 eV at a shock with $M_A \sim 6$ and at a range of θ_{Bn}. Note the power law behaviour of both upstream and downstream populations. Furthermore, compared to simple adiabatic theory, the flux of reflected electrons is relatively high even at angles away from $\theta_{Bn} = 90°$ (considering the range 1–10 keV). These results are in better comparison with the observations, and thus indicate that inclusion of some kind of electron scattering is necessary to fully understand the process of electron acceleration. Scattering by structuring within the shock has shown to be one such viable process.

2.5 Summary

Electron acceleration is a key solar and astrophysical process, and shocks play a role in many situations. We summarize the current status, as informed by our understanding of the Earth's bow shock, as follows: (a) The basic mechanism of fast Fermi acceleration (i.e., adiabatic reflection or shock drift acceleration) is robust and will operate even in the presence of other processes, which may modify it. (b) Curvature has been shown to play a role, and is important if the foreshock electron distributions are to be properly modelled. (c) Full particle simulations have demonstrated many of the qualitative features of observations, but the appropriate application of the results, given the computational constraints of such simulations, is an ongoing research topic. (d) Hybrid simulations have shown that even without electron scale waves the shock can have

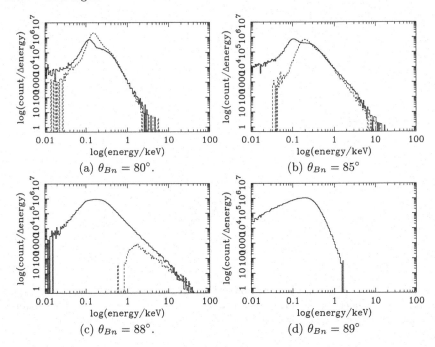

Fig. 6. Differential energy counts (arbitrary units) showing upstream (*dashed*) and downstream (*solid*) final logarithmic electron energy distributions for a fixed Alfvén Mach number $M_A \sim 6$, and for a range of values of θ_{Bn}. (Supplied by R.E. Lowe)

sufficient structure to produce pitch angle scattering of electrons, leading to power law energy distributions both upstream and downstream, in qualitative agreement with observations.

Future work will be to combine some of the different processes affecting electron dynamics, so that their relative importance, and variation with shock parameters, can be evaluated. We will soon be at the stage where better observations will be required in order to distinguish between rival theories.

3 Quasi-parallel Shocks and Acceleration

From the time of the first observations of the bow shock, it has been recognized as a prolific source of energetic ions. The properties of the energetic ions with near-isotropic velocity space distributions, the so-called diffuse ions, have been much studied. The energy range spans from several keV through to ~150 keV, with protons and alpha particles having energy spectra which can be described as exponential in energy per charge [31]. The angular distribution is fairly isotropic, centred near to the solar wind velocity. A crucial characteristic is that the diffuse ions are always associated with ULF (0.01 – 0.2 Hz) magnetic field activity at well above ambient solar wind levels. In time the energetic particles are observed as discrete events. However, they are seen simply when the magnetic field through the observation point connects with a location on the bow shock where the shock geometry is quasi-parallel. Thus, temporal changes reflect the changes in the solar wind magnetic field direction.

The association of wave activity, energetic particles and the appropriate geometry for particle propagation away from the shock: all these indicate that the phenomenon is to be understood as a coupled shock-foreshock system. This will become more evident when we discuss the structure of the quasi-parallel shock itself.

Given the close, one to one, association between energetic particles and magnetic waves, a natural explanation for the acceleration is diffusive, or first-order Fermi acceleration. A simple model is one in which the waves upstream propagate slowly in the plasma frame, but are convected by the supersonic flow towards the shock. Similarly, downstream waves are also convected in the now compressed and slowed shocked flow. The field fluctuations have the effect of perturbing the particle trajectories, and thus act as scattering centres embedded in the flow, or can be considered as such if the particle velocity is greater than the wave speeds relative to the flow. The net effect is that the field fluctuations allow the energetic particles to couple to the flow, and thus experience the plasma compression across the shock. Two different analytic approaches have been developed to treat this problem. The first [32] solves a one dimensional diffusion equation with a given discontinuity in the background flow speed. The second approach [33, 34] adopts an equivalent single particle probabilistic method, which clearly demonstrates how energization results from the particles bouncing back and forth between the scattering centres, which are converging in the shock frame. Particles gain energy at the shock until they escape downstream or upstream.

Imposing the requirement of a time stationary solution on these theories has a number of major consequences, which illustrate the universality and importance of shock acceleration. A power law energy spectrum of the accelerated particles is predicted that is independent of the initial injection spectrum (provided it is steeper than the final spectrum), and independent of the details of the scattering process (provided that the distribution function remains isotropic to first order). Furthermore, the index of the power law of the accelerated spectrum depends only on the shock compression ratio. Since this is limited to a factor four at high Mach number shocks [1], there is a characteristic spectral index for shock accelerated particles. Full details of theories of first-order Fermi acceleration are reviewed in, e.g., [35, 36, 37].

A large body of research exists concerning diffusive shock acceleration, and so we note here some issues which are still unresolved, or where there has been recent progress. The simple test particle approach assumes the scattering centres and the shock discontinuity as given. However, since the scattering is expected to be a resonant cyclotron process, it is natural to assume that the waves are being self-consistently generated by the energetic particles themselves. This has led to quasi-linear theories for diffusive acceleration, in which there is energy flow from the particles into the waves, but the properties of the waves themselves are given by linear theory [38]. So far the definitive test of such theories at the Earth's bow shock has not been carried out, although related work exists, based on observations of interplanetary shocks [39]. At the bow shock, there is evidence that magnetic field fluctuations do not present a power law spectrum (as assumed by the models) over the required frequency range [40]. The upstream waves show evidence of nonlinear behaviour, and the consequences of this for energetic particle propagation and diffusion have not been fully explored.

The simple diffusive acceleration theories assume an initial population which is energized: the so-called injection population. Whether or not such a population could exist, or be extracted from the thermal population, is broadly called the injection problem. The simple theories also assume a shock discontinuity in the flow, without structure or length scales. This, of course, cannot be realistic. A major advance was to use simulations of quasi-parallel shocks to study the coupled processes of thermalization, particle injection and acceleration [41, 42, 43, 44].

These studies have shown the process of particle acceleration to be intrinsic to the structure of the quasi-parallel shock. The picture that has emerged is as follows: The shock layer is highly dynamic, with no single shock surface, but rather a superposition of sharp and weak field gradients and magnetic pulsations which propagate and compress through the shock. The dynamics at the shock layer are strongly influenced by the upstream (foreshock) ULF waves which are convected into the shock. Ion thermalization occurs by partial specular reflection of some particles (as at the quasi-perpendicular shock) or by compression of unshocked plasma between large amplitude magnetic pulsations. The quasi-parallel geometry allows some reflected ions to propagate upstream. Effectively, this is the injection population required for diffusive acceleration. The process of wave scattering and diffusive acceleration can be followed in the simulations and power law energy spectra can be recovered, as well as the exponential density fall off in the upstream region expected from diffusive transport [43].

3.1 Cluster Observations of the Quasi-parallel Pulsation Shock

From observations of high Mach number quasi-parallel shocks it is easy to state that the transition is spatially extended, inhomogeneous and temporally dynamic. With single spacecraft data it is difficult to say any more. Based on theoretical ideas for cosmic ray mediated shocks, a possibility was the existence of a "sub-shock" within a larger scale deceleration region, where the deceleration (and weakening of the sub-shock) was associated with the energetic particle pressure. However, this model never had observational support from bow shock data. Instead large amplitude, highly dynamic magnetic waves and pulsations were observed. Dual spacecraft measurements, combined with results from numerical simulations, revealed that observed changes in the magnetic field strength could not all be explained by simple in-out motion of a shock surface, as, for example, at the quasi-perpendicular shock. Instead there are coherent, short-scale magnetic pulsations embedded in the flow, within the overall transition. In the plasma frame the pulsations are propagating in the upstream direction, but the supersonic flow convection ensures that they are driven into the shock, e.g., [45, 46].

A model was developed [47] in which the magnetic field pulsations, known as SLAMS (for short, large amplitude magnetic structure), are the basic constituent of the shock. The shock itself is composed of a patchwork of such pulsations which slow and pile up at the nominal shock location. Instead of a single sub-shock (as in the model of a cosmic ray mediated shock), the plasma deceleration and thermalization occurs at multiple sites and structures. This model leads to a picture with both spatial inhomogeneity, since the structures have finite transverse extent, and temporal variability, since the shock gradient sharpens and weakens as the pulsations convect through. The pulsation structure also modulates the thermalization and particle injection at the shock. It is expected that the pulsations, with peak amplitudes more than a factor four above the upstream values, grow from individual ULF foreshock waves. As they approach the shock they gain energy from the gradient in the suprathermal particles, growing in amplitude and steepening as they do so [44, 48].

The pulsations also form the basis of the downstream turbulence. This model, although attractive since it is based on simulation work, still leaves open many unresolved issues. For example: what is the overall size and shape of the pulsations transverse to the plasma flow direction? Do they have any internal structure? How do they reflect ions from the thermal distribution? How much of the downstream shocked plasma actually passes through a magnetic pulsation? What is the growth time from ULF wave to pulsation? Some of these questions have been investigated with simulations. For example, the evolution and steepening of pulsations has been examined [49].

Recent studies using Cluster data have also shed some light on the various scale lengths within the quasi-parallel shock. At first sight the four spacecraft data of the Cluster mission offers an ideal opportunity to study scale lengths and time scales. However, the studies have demonstrated that some subtlety is required for interpretation of the data. For example, if variations are seen at only one of the four points, then one can only deduce a scale length lower limit for the edge of a structure, not the structure itself. This has the unfortunate implication that for many studies, unless the observations are carefully interpreted, four points of measurement are not enough. Another difficulty has simply been that Cluster has observed relatively few quasi-parallel shocks, due to a combination of solar wind conditions, the relatively high latitude orbit and, early in the mission, insufficient day-side telemetry coverage. Nevertheless, some progress has been made which is reviewed next; see [50] for a full review of the Cluster view of the quasi-parallel bow shock.

An overview of the nature of the quasi-parallel shock transition is shown in Fig. 7, which presents magnetic field and ion data from Cluster 3 for one

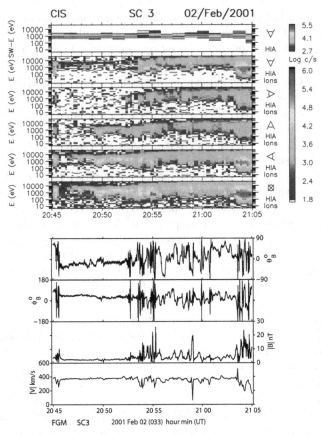

Fig. 7. Extended transition from clean solar wind (*left*) to the magnetosheath (*far right*) showing the appearance and evolution of diffuse ions (*top*) together with magnetic fluctuations at the bow shock under quasi-parallel conditions. The top panel shows data from the CIS HIA instrument. Sub-panels show ion energy flux in counts per second from different directions indicated by the key on the right hand side: sunwards (*up*); dusk (*left*); dawn (*right*); earthward (down). HIA was operating in solar wind mode at this time, so the solar wind beam is measured separately and plotted in the top sub-panel. The bottom sub-panel shows the omnidirectional flux. The bottom panel shows magnetic field and plasma velocity. (From [50], supplied by E. Lucek)

small part of such a transition. In the bottom panels the ion speed, field magnitude and angles are given. The upper panels are, from the top, solar wind thermal ions (the ion instrument resolves these separately from the full 3D distribution), and then ion count rates in the range 10–20 keV in four azimuthal look directions, and finally the omnidirectional count rate. The look directions are indicated on the right hand side of each panel.

Unshocked solar wind can be seen at approximately 20:46 UT, where the field and solar wind speed is undisturbed and there are few, if any, energetic ions. Full thermalization in the magnetosheath can be seen from 21:03 UT, where the flow has been significantly slowed and heated. The transition between these two states is marked by the appearance of energetic ions in range 1–20 keV from 20:47 UT, and which become isotropic ("diffuse") from about 20:55 UT. Within this period the field magnitude displays a number of large amplitude pulsations, up to 5 to 6 times the ambient solar wind value. These pulsations can be isolated, discrete events lasting only a few seconds (e.g., at 20:59 UT), or more complex events with several pulsations apparently run together (e.g., 20:53 - 20:55 UT). The isolated events are surrounded by periods of ULF wave activity, similar to that seen in the ion foreshock when more distant from the bow shock. In both isolated and complex pulsation cases there is partial thermalization of the solar wind, seen in the omnidirectional ion flux and localized decreases of the solar wind speed. Although the figure only shows the edge of the final thermalized zone (after 21:03 UT), the field structure is very similar in amplitude and wave-like properties to the complex pulsation events. Alternatively it may be viewed as a superposition of many of the isolated type of pulsation.

Even from this overview a number of key facts can be drawn: Pulsations occur within periods of diffuse energetic ions, close to the bow shock, and embedded within periods of ULF waves. This is the primary justification for the model of pulsation growth from the ULF wave field, driven by energetic particle gradients close to the shock.

The ion instrument on Cluster has a time resolution of a spin (4 s), which limits what can be done in terms of tracing the ion distributions through the pulsations, since they only last for 10–15 s. Figure 8 shows a time interval of only one minute, and three ion distributions (in the GSE [1] V_x-V_z plane). The magnetic field magnitude for all four Cluster spacecraft is shown in the upper panel. This was a period when the separation scale was 100–200 km, and it can be seen that, on average, all four points measure similar fields. However, in the magnetically "noisy" period around 17:07:05 UT there are appreciable differences between the four measurements, indicating short scale turbulence at high amplitude. Even for the isolated pulsation at 17:07:15 UT, one of the spacecraft measures a magnetic field magnitude which is consistently different from the others, despite the small separations.

The periods when the ion distributions were collected is shown by the grey shaded intervals. The last distribution is typical of the diffuse ion distributions, with a narrow, unthermalized solar wind core. (It is slightly distorted since it includes the alpha particle solar wind population.) The second distribution shows a strong, beam-like, backstreaming component in the upper-left quadrant. It is likely that both ion distributions are time-aliassed to a greater or lesser extent, so identifying ion distribution features with magnetic field changes has to be done carefully, and is best done on the basis of several

---

[1] Geocentric Solar Ecliptic: A geocentric system with x towards the Sun, z towards the Ecliptic North Pole, and y completing the right-handed triad.

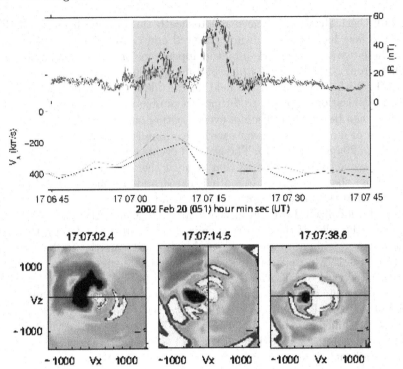

Fig. 8. Ion reflection associated with SLAMS. The top two panels show magnetic field magnitude and plasma velocity in the X_{GSE} direction. The *three shaded areas* show the approximate periods over which the ion distributions shown below were accumulated. Each ion distribution shows ion flux plotted on a colour scale where *blue* and *green* indicate low fluxes, and *black* indicates high flux. Each distribution is a cut through $V_Y = 0$ and shows V_X on the ordinate and V_Z on the abscissa. (From [50], supplied by E. Lucek)

similar events. Finally, the first distribution, during the period of strong short scale turbulence, shows a slowed and deflected solar wind and a narrow component (on positive V_z axis) which would be consistent with specular reflection of a fraction of the solar wind. In the picture of the quasi-parallel pulsation shock, what is revealed here is the moment when solar wind ions are being extracted from the thermal distribution, and boosted in energy so that they can end up as the Fermi accelerated diffuse ions. More work remains to be done to understand the details of the reflection processes (probably similar to the reflected-gyrating particles at the quasi-perpendicular shock) and the interaction with the solar wind (a variety of ion beam instability). However, the importance of quasi-parallel shock structure for the injection of energetic ions is beyond dispute.

3.2 Diffusive Transport at Quasi-parallel Shocks

As discussed above, first-order Fermi acceleration theories can be used successfully to explain the acceleration of ions at the quasi-parallel shock. However, detailed tests of all aspects of the theories have yet to be completed. The key physical assumption underlying such theories is that the particles undergo diffusive transport in the upstream region. Steady state theories then predict an exponential fall-off in the energetic particle density ahead of the shock. From the e-folding distance $L(E)$, as a function of energy E, the spatial diffusion coefficient and hence the mean free path can be derived. Previous studies relied

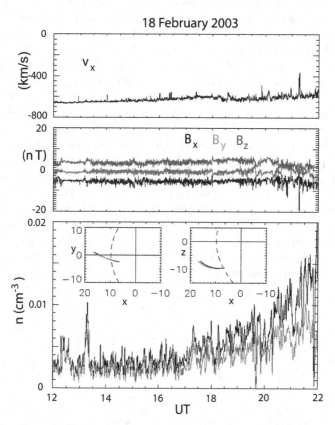

Fig. 9. Data from which the spatial gradient in diffuse ions, and hence the scattering mean free path, have been determined. **From top to bottom**: Solar wind velocity component vx and magnetic field components Bx (*black line/lower curve*), By (*blue line/central curve*), Bz (*red line/upper curve*) as measured on Cluster 1, partial ion density in the 24 - 32 keV energy range as measured at Cluster 1 (*black line/upper curve*) and Cluster 3 (*green line/lower curve*). Also shown in the lower panel are projections of the spacecraft orbits and bow shock onto the $x - y$ and $x - z$ plane, respectively. (From [52], supplied by M. Scholer)

on single spacecraft data, which meant that, given the ever-changing state of the solar wind, only statistical studies of the upstream energetic particle density profile could be carried out. For example, an e-folding distance which varied from $3.2 \pm 0.2\, R_E$ at 10 keV to $9.3 \pm 1.0\, R_E$ at $\sim$67 keV was found [51]. However, the statistical fit did not have a very high correlation coefficient, and so provided some room for doubt.

Recently Kis et al. [52] have used Cluster data to give direct evidence of the exponential fall-off of energetic particle density. Data was used from a time in the mission when the inter-spacecraft separation was relatively large (about $1.5\, R_E$), and when the solar wind conditions stayed relatively constant so that the foreshock was sampled over a period of several hours. An overview of the data is shown in Fig. 9, which demonstrates how the partial density of the energetic ions in the energy range 24–32 keV increases towards the shock (on the right hand side of the figure). It is also apparent that there is considerable variability, which is believed to be due to changes in the location of the bow shock in response to changes in the solar wind state. Using a model for the bow shock location, an absolute distance (from the shock) can be inferred. Then, using the relative density at two of the Cluster spacecraft, the log density gradient as a function of distance from the shock can be plotted (Fig. 10). The linear relationship in that figure confirms that the density profile is exponential as a function of distance from the shock. An e-folding distance of $\sim$0.5R_E at 11 keV to $\sim$2.8 R_E at 27 keV was found. It is not clear why these values are smaller than those found earlier [51]. Further work will allow comparison with the magnetic field spectra and the predictions of quasi-linear theory. However, this work has demonstrated in a very direct manner that upstream particles undergo diffusive transport in the upstream region. This can be viewed as direct evidence for the underlying physical mechanism of first-order Fermi acceleration.

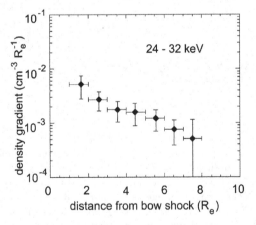

Fig. 10. Average partial ion density gradient in the 24–32 keV energy range versus distance from the bow shock. (From [52], supplied by M. Scholer)

4 Transients at Shocks: Hot Flow Anomalies

In searches of data for bow shock crossings a peculiar signature was sometimes noted, in which the solar wind flow was strongly deflected (sometimes by more than 90° from its usual direction) and heated. Studies of these events eventually led to their recognition as a distinct class, now called "hot flow anomalies" (HFA). Initially identified in ISEE and AMPTE data [53, 54], they occur upstream of the bow shock and can appear as a region of depressed field strength filled with hot plasma. Clearly, a volume of hot, magnetosheath-like plasma surrounded by ambient solar wind represents a significant perturbation, and HFA's have been seen to drive shocks on their outer edges.

A typical example of an HFA as observed by Cluster is shown in Fig. 11 [55]. In the region marked by the second heavy bar the magnetic field becomes low, where the proton density is also low, but with an elevated temperature, as in a diamagnetic cavity. There is considerable magnetic wave activity in the low field region. This is followed by an increase of both field and density, which is interpreted as being caused by a compression driven by the hot low density plasma. Almost immediately afterwards, apparently separated by a brief period of solar wind, comes a second field and plasma compression. For

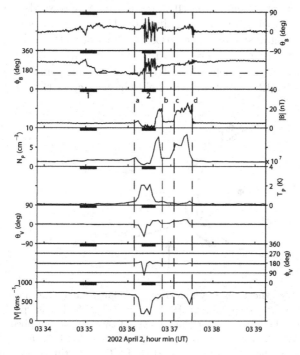

Fig. 11. Cluster measurements of HFA event showing magnetic field (angles and magnitude), plasma density, temperature and velocity (angles and magnitude). (From [55]. Supplied by E. Lucek)

this particular event the second compression region is explained by the HFA further expanding to pass over the spacecraft and then finally receding again.

The properties of HFA's can be summarized as follows [56]: They contain regions of hot ($T \sim 10^6$–10^7K) plasma with near isotropic ion and electron distributions. The flow in the events is slower than the ambient solar wind, and strongly deflected. There can be regions, usually in the centre of the events, with strong field perturbations, both depressions and enhancements. Compression regions are sometimes observed on the edges, with the outer boundaries consistent with being weak shocks. Crucial to understanding their formation is that they are observed close to the bow shock and that they are almost always associated with the presence of an interplanetary current sheet, i.e., directional discontinuity of some kind.

The currently favoured model for HFA formation is that they are produced when a thin tangential discontinuity (TD) in the solar wind interacts with the bow shock. Simulations [57, 58, 59] indicate that the disruption of a quasi-perpendicular shock by a TD can produce a channel of hot magnetosheath-like plasma which extends into the upstream region along the TD. The expansion of this plasma produces the low field regions and flow deflection, and also explains the compression of the ambient solar wind on its edges. From the modelling, it was found that the orientation of the TD has to be such that the motional electric field on both sides is directed towards the TD, and this has been confirmed observationally [56]. This configuration ensures that particles streaming upstream are focussed back towards the TD, which guides them into the upstream and away from the shock, thus disrupting the shock. The importance of the orientation of the TD and motional electric field is illustrated in Fig. 12. Essentially the field change across the TD, relative to

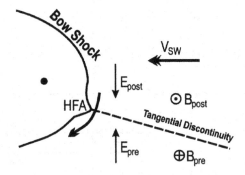

Fig. 12. Sketch of the interaction between a solar wind tangential discontinuity (TD) and the bow shock creating an HFA (after [56]). The labels "pre" and "post" refer to fields seen before and after the passage of the TD over an observation point close to the HFA. The electric field is the motional electric field $\boldsymbol{E} = -\boldsymbol{V}_{SW} \times \boldsymbol{B}$, and points towards the TD, in the sense that the $\boldsymbol{E} \times \boldsymbol{B}$ drift is along the TD and away from the bow shock. The motion of the HFA around the bow shock, as the TD is convected by the solar wind, is indicated by an arrow

the solar wind direction, has to be such that the motional field points towards the TD. The effect is that the $\boldsymbol{E} \times \boldsymbol{B}$ drift of the particles takes them along the TD, away from the shock. The orientation of the TD relative to the curved bow shock determines how fast the HFA is dragged over the shock, and it is found that this needs to be relatively slow in order for an HFA to develop [56]. Note that in the solar wind tangential discontinuities are expected to have an isotropic distribution of orientations. TD's with an orientation edge-on to the bow shock will spend the longest time interacting, and so will have the strongest signatures.

In terms of energetic particles, we must address the question whether HFA's are of any general importance, or merely a minor perturbation of the bow shock's surface. Suprathermal particles are associated with HFA's, as seen in Fig. 13 which shows two cuts through the ion distribution before and after entering the region of heated, deflected flow. At times, particles are seen with energies comparable to the diffusively accelerated ions. Clearly the energies of these ions are not spectacularly high. But one aspect of HFA's does appear significant. Quasi-perpendicular shocks are highly efficient at heating ions, but the shock geometry ensures that all but the most energetic ions are transported downstream. However, the HFA mechanism is one which lead to injection of heated and accelerated ions into the region upstream of the shock. At the terrestrial bow shock the full consequences of this are not seen, since the magnetic field geometry changes rapidly over the shock surface. But one might suppose, in a situation where the quasi-perpendicular shock is planar, that the ions injected at the HFA might eventually scatter in the upstream region and return to the shock as an injection population for some further acceleration process. The condition for this to operate efficiently would be that a quasi-perpendicular shock should propagate through a region where there were sufficient TD's or current sheets. There are several situations where this

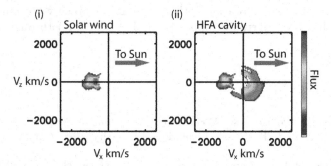

Fig. 13. Two cuts through Cluster 3 CIS ion distributions, in the $X - Z$ GSE plane, showing (i) the ion population before the spacecraft entered the HFA and (ii) the heated ion population flowing sunward and southward within the HFA. The log of the particle flux is indicated by the color scale, with *red* and *orange* regions indicating higher flux levels. (From [55], supplied by E. Lucek)

is likely, such as at interplanetary shocks in the heliosphere, or at shocks propagating in the corona, driven by, e.g., coronal mass ejections or flares. So far most work has concentrated on the possibilities of ion acceleration at HFA's. However, heated electron distributions are observed within HFA's, and it represents an interesting avenue to study whether efficient injection, and subsequent acceleration, of electrons, as well as ions, can occur at quasi-perpendicular shocks due to the interaction with upstream discontinuities.

5 Summary

Observations of the Earth's bow shock continues to produce new results on particle acceleration, for both electrons and ions. For studying the process of energizing thermal particles, the bow shock is probably ideal, and results here have implications for particle acceleration in the solar and astrophysical context. The bow shock is also a complex environment, in terms of geometry and variability, and so can lead to some productive analogies which can be used to model other phenomena in the heliosphere. There are limitations with this approach, and so it is important to understand how the general processes work in the context of the bow shock, in order to apply them to, for example, coronal shocks. Nevertheless, there are still many discoveries to be made, and by combining our knowledge of the bow shock and coronal shock acceleration we can hope to reach a more complete understanding of both.

Acknowledgements

Thanks are given to members of the ISSI *Dayside Magnetospheric Boundaries* Working Group for useful discussions and figures. Some of the work presented was carried out in collaboration with Robert Lowe. Some work reported in this paper has been partially supported by PPARC (UK) through research grants.

References

1. D. Burgess: "Collisionless shocks", in *Introduction to Space Physics*, ed. by M.G. Kivelson, C.T. Russell (Cambridge University Press, Cambridge, 1995), p. 129
2. S.D. Bale, M.A. Balikhin, T.S. Horbury, V.V. Krasnoselskikh, H. Kucharek, E. Möbius, S.N. Walker, A. Balogh, D. Burgess, B. Lembège, E.A. Lucek, M. Scholer, S.J. Schwartz, M.F. Thomsen: Space Sci. Rev. **118**, 161–203 (2005)
3. P.C. Filbert, P.J. Kellogg: J. Geophys. Res. **84**, 1369–1381 (1979)
4. I.H. Cairns: J. Geophys. Res. **92**(11), 2315–2327 (1987)
5. R.J. Fitzenreiter, J.D. Scudder, A.J. Klimas: J. Geophys. Res. **95**, 4155–4173 (1990)

6. D. Burgess: Adv. Space Res. **20**(4/5), 673–682 (1997)
7. S.A. Knock, I.H. Cairns, P.A. Robinson, Z. Kuncic: J. Geophys. Res. **108**(A3), 6–1 (2003)
8. J.J. Mitchell, I.H. Cairns, P.A. Robinson: J. Geophys. Res. **109**(A18), 6108 (2004)
9. Z. Kuncic, I.H. Cairns, S.A. Knock: J. Geophys. Res. **109**(A18), A02108 (2004)
10. K.A. Anderson, R.P. Lin, F. Martel, C.S. Lin, G.K. Parks, H. Rème: Geophys. Res. Lett. **6**, 401–404 (1979)
11. M. Vandas: J. Geophys. Res. **106**(A15), 1859–1872 (2001)
12. W.C. Feldman, R.C. Anderson, S.J. Bame, S.P. Gary, J.T. Gosling, D.J. McComas, M.F. Thomsen: J. Geophys. Res. **88**(NA1), 96–110 (1983)
13. R.J. Fitzenreiter, A.J. Klimas, J.D. Scudder: Geophys. Res. Lett. **11**, 496–499 (1984)
14. J.T. Gosling, M.F. Thomsen, S.J. Bame, C.T. Russell: J. Geophys. Res. **94**(NA8), 10011–10025 (1989)
15. M.M. Leroy, A. Mangeney: Ann. Geophys. **2**(4), 449–456 (1984)
16. C.S. Wu: J. Geophys. Res. **89**(NA10), 8857–8862 (1984)
17. R.J. Fitzenreiter: Adv. Space Res. **15**(8/9), 9–27 (1995)
18. I.H. Cairns, P.A. Robinson, R.R. Anderson, R.J. Strangeway: J. Geophys. Res. **102**(11), 24249–24264 (1997)
19. D. Krauss-Varban, D. Burgess, C.S. Wu: J. Geophys. Res. **94**(11A), 15089–15098 (1989)
20. D. Krauss-Varban, D. Burgess: J. Geophys. Res. **96**(A1), 143–154 (1991)
21. M. Vandas: J. Geophys. Res. **100**(A9), 23499–23506 (1995)
22. C.F. Kennel, J.P. Edmiston, T. Hada: "A quarter century of collisionless shock research", in *Collisionless Shocks in the Heliosphere: A Tutorial Review, Geophys. Monogr. Ser.*, Vol. 34, ed. by R.G. Stone, B.T. Tsurutani (AGU, Washington, D. C., 1985), pp. 1–36
23. V.V. Krasnoselskikh, B. Lembège, P. Savoini, V.V. Lobzin: Physics of Plasmas **9**, 1192–1209 doi:10.1063/1.1457465 (2002)
24. B. Lembège, P. Savoini: Physics of Fluids B **4**, 3533–3548 (1992)
25. M. Scholer, S. Matsukiyo: Ann. Geophys. **22**, 2345–2353 (2004)
26. B. Lembège, P. Savoini: J. Geophys. Res. **107**(A3) (2002)
27. P. Savoini, B. Lembège: J. Geophys. Res. **106**(A15), 12975–12992 (2001)
28. D. Krauss-Varban: J. Geophys. Res. **99**(A18), 2537–2551 (1994)
29. R.E. Lowe, D. Burgess: Geophys. Res. Lett. **27**, 3249–3252 (2000)
30. R.E. Lowe, D. Burgess: Ann. Geophys. **21**, 671–679 (2003)
31. F.M. Ipavich, A.B. Galvin, G. Gloeckler, M. Scholer, D. Hovestadt: J. Geophys. Res. **86**(NA6), 4337–4342 (1981)
32. W.I. Axford, E. Leer, G. Skadron: "The acceleration of cosmic rays by shock waves", in *Proc. Int. Conf. Cosmic Rays 15th* (Vol. 11, 1977), p. 132
33. A.R. Bell: M. Not. R. Astron. Soc. **182**, 147–156 (1978)
34. A.R. Bell: M. Not. R. Astron. Soc. **182**, 443–455 (1978)
35. M. Scholer: "Diffusive acceleration", in *Collisionless Shocks in the Heliosphere: Reviews of Current Research, Geophys. Monogr. Ser.*, Vol. 35, ed. by R.G. Stone, B.T. Tsurutani (AGU, Washington, D. C., 1985), pp. 287–301
36. M.A. Forman, G.M. Webb: "Acceleration of energetic particles", in *Collisionless Shocks in the Heliosphere: Reviews of Current Research, Geophys. Monogr. Ser.*, Vol. 35, ed. by R.G. Stone, B.T. Tsurutani (AGU, Washington, D. C., 1985), pp. 91–114

37. F.C. Jones, D.C. Ellison: Space Sci. Rev. **58**, 259–346 (1991)
38. M.A. Lee: J. Geophys. Res. **87**, 5063–5080 (1982)
39. C.F. Kennel, F.V. Coroniti, F.L. Scarf, W.A. Livesey, C.T. Russell, E.J. Smith, K.P. Wenzel, M. Scholer: J. Geophys. Res. **91**, 11,917–11,928 (1987)
40. T. Terasawa: Adv. Space Res. **15**(8/9), 53–62 (1995)
41. D. Burgess: Geophys. Res. Lett. **16**, 345 (1989)
42. V.A. Thomas, D. Winske, N. Omidi: J. Geophys. Res. **95**, 18,809–18,819 (1990)
43. J. Giacalone, D. Burgess, S.J. Schwartz, D.C. Ellison: Ap. J. **402**, 550–559 (1993)
44. M. Scholer, D. Burgess: J. Geophys. Res. **97**(A6), 8319–8326 (1992)
45. M.F. Thomsen, J.T. Gosling, C.T. Russell: Adv. Space Res. **8**(9/10), 9175–9178 (1988)
46. S.J. Schwartz, D. Burgess, W.P. Wilkinson, R.L. Kessel, M. Dunlop, H. Lühr: J. Geophys. Res. **97**, 4209–4227 (1992)
47. S.J. Schwartz, D. Burgess: Geophys. Res. Lett. **18**, 373–376 (1991)
48. J. Giacalone, S.J. Schwartz, D. Burgess: Geophys. Res. Lett. **20**, 149–152 (1993)
49. K. Tsubouchi, B. Lembège: J. Geophys. Res. **109**, H02114, doi: 10.1029/2003JA010014 (2004)
50. D. Burgess, E.A. Lucek, M. Scholer, S.D. Bale, M.A. Balikhin, A. Balogh, T.S. Horbury, V.V. Krasnoselskikh, H. Kucharek, B. Lembège, E. Möbius, S.J. Schwartz, M.F. Thomsen, S.N. Walker: Space Sci. Rev. **118**, 205–222 (2005)
51. K.J. Trattner, E. Möbius, M. Scholer, B. Klecker, M. Hilchenbach, H. Luehr: J. Geophys. Res. **99**, 13389–13400 (1994)
52. A. Kis, M. Scholer, B. Klecker, E. Möbius, E.A. Lucek, H. Rème, J.M. Bosqued, L.M. Kistler, H. Kucharek: Geophys. Res. Lett. **31**, 20801 (2004)
53. S.J. Schwartz, C.P. Chaloner, P.J. Christiansen, A.J. Coates, D.S. Hall, A.D. Johnstone, M.P. Gough, A.J. Norris, R.P. Rijnbeek, D.J. Southwood, L.J.C. Woolliscroft: Nature **318**, 269–271 (1985)
54. M.F. Thomsen, J.T. Gosling, S.A. Fuselier, S.J. Bame, C.T. Russell: J. Geophys. Res. **91**, 2961–2973 (1986)
55. E.A. Lucek, T.S. Horbury, A. Balogh, I. Dandouras, H. Rème: J. Geophys. Res. **109**(A18), 6207 (2004)
56. S.J. Schwartz, G. Paschmann, N. Sckopke, T.M. Bauer, M. Dunlop, A.N. Fazakerley, M.F. Thomsen: J. Geophys. Res. **105**(A14), 12639–12650 (2000)
57. D. Burgess: J. Geophys. Res. **94**, 472 (1989)
58. V.A. Thomas, D. Winske, M.F. Thomsen, T.G. Onsager: J. Geophys. Res. **96**(NA7), 11625–11632 (1991)
59. Y. Lin: Planet. Space Sci. **50**, 577–591 (2002)

On the Existence of Non-maxwellian Velocity Distribution Functions in the Corona and their Consequences for the Solar Wind Acceleration

Milan Maksimovic

LESIA & CNRS, Observatoire de Paris, 92195 Meudon cedex, France
milan.maksimovic@obspm.fr

Abstract. Non-thermal electron and ion velocity distribution functions are permanently observed in the solar wind. The exact origins of such departures from equilibrium Maxwell-Boltzmann distributions remain unclear. It is however believed that the rarity of Coulomb collisions in most of the extended corona and solar wind plays a crucial role in the mechanisms which produce and/or maintain such distributions.

In this paper we focuss more on the electron distribution functions. We summarize their various observations and discuss about their possible coronal origin and role in the Solar Wind acceleration processes.

1 Introduction

An important assumption inherent to the usual fluid solar wind models is that the plasma is at equilibrium, dominated by collisions. Therefore the hydrodynamic approach implies that the particle velocity distribution functions are rather close to a Maxwellian. However the observed solar wind electron and ion distributions depart from nearly isotropic Maxwellians, indicating the limited validity of this hypothesis. As a consequence there is a high level of complexity inherent to the fluid modelisation of the solar wind. In these models a complicated set of transport equations can be obtained by taking higher moments of the Boltzmann equation and "closing" the system by assuming a specific parameterized form for the velocity distribution functions in order to take into account their non-thermal character [2].

Solar wind proton distributions have anisotropic cores that are well represented by bi-Maxwellian distributions aligned to the local magnetic field, with $T_\perp > T_\parallel$ [20]. The protons also exhibit an additional field-aligned "beam" component that flows ahead of the core by about the local Alfvén speed. This beam is mainly present in the fast solar wind.

Contrary to the protons, the observed electron velocity distribution functions (eVDFs) permanently exhibit non-Maxwellian features whatever is the

M. Maksimovic: *On the Existence of Non-maxwellian Velocity Distribution Functions in the Corona and their Consequences for the Solar Wind Acceleration*, Lect. Notes Phys. **725**, 191–202 (2007)
DOI 10.1007/978-3-540-71570-2_9

type of wind, slow or fast, in which they are observed. The eVDFs permanently exhibit three different components: a thermal core and a supra-thermal halo, which are always present at all pitch angles, and a sharply magnetic field aligned "strahl" which is usually antisunward-moving [5, 24, 27, 29]. Since in most of the kinetic and exospheric models, reviewed in this paper, the solar wind electrons play a major role, we will mainly restrict to this particle species in the following.

What is the origin of the non-Maxwellian eVDFs observed in the solar wind? Are such distributions already present in the solar corona or are they only a consequence of the solar wind transport in the interplanetary medium? There is an increasing amount of both theoretical [3, 8, 14, 36, 37] and observational [1, 4, 10, 28] evidences that tend to show that nonthermal VDFs can develop and exist in the high corona and even in the transition region. This is because, in a plasma, the particle mean free paths increase rapidly with speed ($\propto v^4$), so that high-energy tails can develop for Knudsen numbers as low as 10^{-3} [34] that is, even in a semi-collisional plasma. Even more, high-energy tails can be expected to be found in the weakly collisional corona and solar wind acceleration region. However, until now, there have not been any conclusive observations that have settled the question of the shape of the eVDFs in the corona.

In Sect. 2 we review the principal features of the observed eVDFs in the solar wind and discuss the various models that have been used to fit them. In Sect. 3 we review the different coronal and solar wind models where the presence of non-thermal VDFs in the medium has been assumed. This includes kinetic models of the transition region and corona and kinetic/exospheric models that quantify the consequences for the solar wind acceleration of nonthermal eVDFs as boundary conditions in the corona. Finally in Sect. 4 we give some concluding remarks.

2 The Principal Features of Observed Solar Wind Electron Distribution Functions

As we said in the introduction, the observed eVDFs exhibit permanent strong departures from Maxwellian distribution. These departures are usually attributed to the rarity of Coulomb collisions in the corona and solar wind. In order to quantify the effect of collisions it is useful to compute the Knudsen number K_n which is the ratio between the Coulomb collision mean free path (m.f.p.) and a typical hydrostatic scale height. For instance, if the density scale height is used, the Knudsen number is defined as:

$$K_n = \frac{\lambda}{H} \quad \text{where} \quad \lambda = 9.2\ 10^7 \frac{T_e^2}{N_e}(\text{S.I.}) \quad \text{and} \quad H = \left(-\frac{d\ln N_e}{dr}\right)^{-1} \quad (1)$$

where N_e and T_e are respectively the electron density and temperature. The Coulomb mean free path in (1) is that of electron-electron collisions. For the protons, the Knudsen number is comparable since the proton-proton Coulomb m.f.p. is close to the electron-electron one. In Fig. 1 we display, as a function of the distance from the surface in solar radii, the variations of N_e (light/red line) and T_e (heavy/black line). This Figure is typical for a coronal hole structure and its extension into a fast solar wind. The density and temperature curves are obtained by interpolation between various observations which are described in the figure caption. The transition region, at around 2000 km or $3 \ 10^{-3} \ R_s$ of height between the chromosphere and the low corona corresponds to the abrupt increase of T_e from $\approx 10^4$ K to about $5 \ 10^5$ K. From these radial profiles of N_e and T_e, the Knudsen number is displayed in Fig. 2.

A noticeable feature on this Figure are the 13 orders of magnitude of difference between the chromosphere and the corona for K_n. This is a huge difference which unavoidably has to imply the use of different physical approaches in order to model the plasma in these two regions. In the corona K_n ranges between 10^{-2} and 10 and becomes roughly constant at large heliocentric distance. Therefore the corona and solar wind are plasmas which are neither highly collisional $K_n \ll 1$ nor fully collisionless $K_n \gg 1$. The physics one has to use in order to model this region is somehow between two well defined theoretical approaches, the fluid theory on one hand and the Vlasov approximation on the other. This statement is probably the reason why the problem of the solar wind acceleration is still in debate. As noted by [34], for Knudsen numbers as low as 10^{-3}, distributions with high-energy tails may develop in the medium and the Spitzer-Härm theory [35] is not anymore valid. Therefore the corona is already collisionless just after the transition region. The above conclusions and remarks are also valid for a streamer-like corona where, compared to a coronal hole, the temperature is slightly increased, the

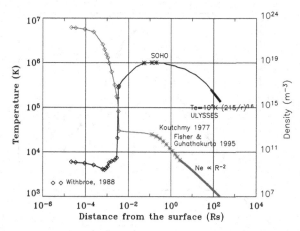

Fig. 1. Variations of the coronal electron density N_e (*light/red line*) and temperature T_e (*heavy/black line*). The density and temperature curves are obtained by interpolation between coronal [6, 11, 39], SoHo [38] and Ulysses [7, 17] observations. This Figure is typical for a coronal hole structure and its extension into a fast solar wind

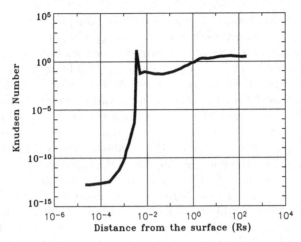

Fig. 2. The radial variation of K_n as a function of height above the Sun's surface

density is roughly larger by one order of magnitude and the typical scale height is almost the same. The Knudsen number being proportional to T_e^2/N_e, its global radial variation remains basically the same, but its value is slightly smaller than in the coronal hole case.

Direct in-situ measurements of VDFs in the corona have still not been performed up to now. In the solar wind on the contrary, a numerous fleet of spacecraft has measured the electron and various ion VDFs. Figure 3 shows a typical eVDF measured in the fast solar wind ($V_{SW} > 650$ km/s) by the Helios spacecraft at a heliospheric radial distance of 0.35 AU [19]. The diamonds and

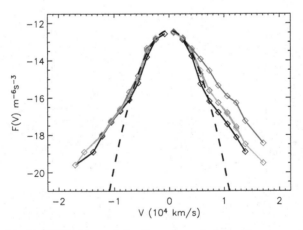

Fig. 3. A typical eVDF observed in the solar wind. In the present case it is a distribution measured by the Helios spacecraft. The diamonds and curves represent different cuts of the eVDF as a function of speed for different values of the pitch angle θ defined in the text

the corresponding colored lines represent different cuts of the eVDF, plotted as a function of the electron speed, for different directions with respect to the direction of the interplanetary magnetic field (IMF). The blue diamonds (upper curve) represent the parallel cut $f^{\parallel}$ along the magnetic field, with the $v > 0$ direction corresponding to the anti-Sunward direction. The black diamonds (lower solid curve) represent the cut $f^{\perp}$ which is perpendicular to the magnetic field. Finally the red and green diamonds are the cuts along directions making angles respectively of 45° (f^*) and 315° (f^{**}) with the IMF in the $(v_{\parallel}, v_{\perp})$ plane. To a good approximation, the distribution is gyrotropic along the IMF, that is symmetric along the IMF in the $(v_{\parallel}, v_{\perp})$ plane. For $v < 0$ all the cuts are roughly equal and for $v > 0$ $f^* = f^{**}$. The black dashed line represents an isotropic Maxwellian VDF f_{Max} which is computed with the values for the density and temperature obtained by integrating respectively the zero and second order moments on the whole observed distribution.

The three different components of the eVDFs in the solar wind can be easily seen on this Figure. An isotropic thermal core is visible for velocities up to roughly $6 \cdot 10^4$ km/s. For larger velocities the VDF departs from f_{Max}. A supra-thermal halo, which is also present at all pitch angles appears. Finally, aligned with the magnetic field, an excess of flux by about a factor of 10 is visible on $f^{\parallel}$. This excess represents the so-called "strahl" component, which is also slightly present on f^* and f^{**} for $v > 0$.

Several models have been used in the past to fit the eVDFs in the solar wind. The classic model is the sum of two bi-maxwellians [5], one for the core population and one for the halo. Other models have been used, which take into account the fact that the flux at high velocities varies more like a power law rather than a Maxwellian. For instance, in [16] Ulysses eVDFs were modelled as generalized Lorentzians or Kappa functions which are defined as:

$$f^{\kappa}(v) = \frac{N_e}{2\pi(\kappa V_{\kappa}^2)^{3/2}} \frac{\Gamma(\kappa+1)}{\Gamma(3/2)\Gamma(\kappa-1/2)} \left(1 + \frac{v^2}{\kappa V_{\kappa}^2}\right)^{-(\kappa+1)}$$

where $\Gamma(x)$ is the Gamma function and V_k is an equivalent thermal speed, related to the equivalent temperature $T_{\kappa} = m < v^2 > /3k$ by

$$V_{\kappa} = \left(\frac{2\kappa - 3}{\kappa} \frac{k_B T_{\kappa}}{me}\right)^{1/2},$$

The Kappa distribution decreases with the speed v as a power law, $f \propto v^{-2\kappa}$. In the limit $\kappa \to \infty$, it reduces to a Maxwellian distribution.

Figure 4 reproduces the $f^{\perp}$ cut of the distribution displayed on Fig. 3. The dashed red and blue lines represent the classical double-Maxwellian core/halo fit. We have fitted the observed VDF with a Kappa model. The result is represented by the solid green line. The best fit to the distribution is obtained for $\kappa = 4.8$. This is consistent with the result obtained by [16] who modelled a large number of observed eVDFs and obtained a parameter κ ranging roughly between 2 and 5.

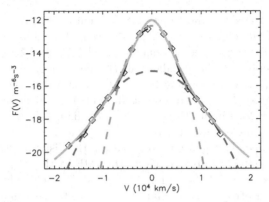

Fig. 4. Perpendicular $f^\perp$ cut of the distribution displayed on Fig. 3. The *dashed red and blue lines* represent the classical double-Maxwellian core/halo fit. The *solid green* represents a fit with a Kappa model, with a parameter $\kappa = 4.8$

As we have seen, the Kappa function is a very common way of representing distributions with non-thermal tails. It is a more economic alternative to the Maxwellian core/halo model for the solar wind electrons since it requires less parameters to be defined (in the isotropic case: one density, one temperature and one parameter κ instead of two densities and two temperatures). However, some studies try to give a more physical ground to such distribution by invoking some new nonextensive entropy formalism [14].

3 Modelling the Consequence of Non-thermal Distributions in the Corona

Most of the models dealing with the physics of the solar corona assume that the particles VDFs are Maxwellian. This is intrinsically the case for the majority of solar wind fluid models as for instance Parker's model [26]. This is usually also the case for most of the theories dealing with the emission and absorption of electromagnetic waves by the coronal plasma and which are used to probe remotely the coronal properties.

However, as we noted in Sect. 2, the corona is highly collisionless just after the transition region and non-thermal distributions could develop there. What will be the consequences if such non-thermal VDFs really exist in the corona and transition region? Several kinetic models have investigated this possibility. We give in this section a brief overview of their main results.

3.1 The Velocity Filtration Mechanism

In 1992, Jack Scudder proposed a new mechanism to explain the high coronal temperature [32, 33]. This mechanism, named "the velocity filtration" effect is based on the assumption that the ion and electron VDFs in the chromo-sphere, and therefore in the corona, are non-Maxwellian, for instance Kappa functions.

Figure 5 presents the graphical demonstration of the velocity filtration. If we assume that collisions and waves can be neglected, then the radial evolution of the particles VDFs is simply given by Liouville's theorem. Let us then assume that the VDFs are Maxwellian. This case is illustrated on the left hand side of the Figure. When plotted in a frame $\ln(f)$ as a function of the energy $E = mv^2/2$ times the sign of v, Maxwellians are represented by straight lines. At an altitude r_0 in the chromosphere the total, gravitational plus electric, potential is Φ_O and the distribution is $f_0(v^2, r_0)$. At an altitude r in the corona the potential is $\Phi_O + \Delta\Phi$ and the distribution $f(v^2, r)$. Close to the sun the electric potential ϕ_E is well approximated by the Pannekoek-Rosseland [25, 31] one: $e\phi_E \approx m_p M_s G/2r$, so that the total potential is attractive for both protons and electrons. From Liouville's theorem, only those particles, represented by the shaded area of the VDF in the bottom of panel (a) and having a total energy larger than $\Delta\Phi$, can reach r. The distribution at r is therefore simply given by $f(v^2, r) = f_0(v^2, r_0) \times \exp(-e\Delta\Phi/k_B T)$. The slope of f and f_0 and therefore their temperatures is unchanged with altitude. Only the density decreases by a factor $\exp(-e\Delta\Phi/k_b T)$ [21].

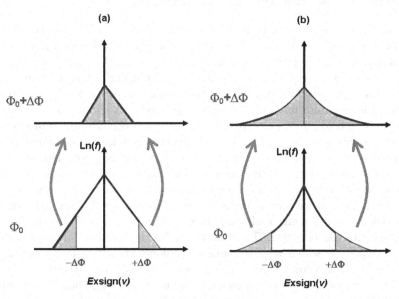

Fig. 5. Graphical demonstration of the "velocity filtration" mechanism. The loga-rithm of the distribution function is plotted as a function of the energy $E = mv^2/2$ of the particles times the sign of v. The panels (**a**) present the evolution of a Maxwellian distribution function in an attractive potential Φ. The panels (**b**) represent the same evolution for a Kappa function

What happens now if we assume a Kappa distribution in the chromosphere? This is illustrated on the right side of Fig. 5. As for the Maxwellian case, only those particles represented by the dashed area of the VDF in the bottom of panel (b) and having a total energy larger than $\Delta\Phi$, can reach r. The distribution at r is given by:

$$f(v^2, r) = f_0 \left(\frac{v^2}{1 + \frac{\Delta\Phi}{(2\kappa - 3/2)k_B T}}, r_0 \right) \times \left(1 + \frac{\Delta\Phi}{(2\kappa - 3/2)k_B T} \right)$$

Contrary to the Maxwellian case, the slope of the Kappa distribution is no more constant with both the energy E of the particles and the altitude. The slope decreases with altitude and the temperature therefore increases. In Scudder's velocity filtration mechanism, there is no need of additional energy to heat the corona. The corona is hot only because the most energetic particles can escape from the chromosphere. By way of generalizing, [22] and [23] have demonstrated analytically the velocity filtration in the special case of a superposition of Maxwellians. They showed that the velocity filtration is a general mechanism and not only an artefact due to the use of Kappa functions.

3.2 Kinetic Models of the Transition Region

Although the corona is a highly collisionless medium, collisions cannot be completely neglected. What will happen to the velocity filtration mechanism if collisions are taken into account properly? Can non-thermal VDFs survive in this case? [8] and [3] have studied this point using different approaches. [8] address this question through kinetic simulations of the low solar corona. They assume a finite number of protons and electrons plunged in a constant gravitational field and an electric field which is self consistently computed in order to ensure quasi-neutrality everywhere in the system. In their simulations, when two particles meet they may make an elastic collision depending on the magnitude of their relative velocity. The functional form of the velocity-dependent collision probability is specific to Coulomb binary collisions. In their work, [3] obtained numerical solutions of the Fokker-Planck equation for electrons in a one dimensional, static layer of hydrogen plasma immersed in a constant external force field. The authors use for the Fokker-Planck equation a collision operator which is also specific to Coulomb binary collisions [30]. Both studies assume a coronal layer which corresponds roughly to the temperature transition region (TR). The interesting point concerning these two studies is that, starting from different approaches of the same problem, they reach the same major conclusions. Firstly if an electron VDF with supra-thermal tails at the base of the TR is imposed, then the collisions are not sufficiently numerous to thermalize completely the plasma at the top of the TR. Non-thermal tails still exist at this altitude, even with a weak Knudsen number (10^{-2} to 10^{-4}) in the TR. Secondly, if one assumes that the electron VDFs at the base of the TR have sufficiently strong suprathermal power law tails, the heat flux may flow upwards, i.e. in the direction of increasing temperature. In other words, when a temperature gradient exists in the layer, nonlocal heat flow responds to the external force in such a way that the electron heat flux can become decoupled from the local temperature gradient. This result is illustrated in Fig. 6. Both [8] and [3] use Kappa VDFs as prototypes for non thermal velocity distributions. On this figure are plotted the values of the normalized total heat flux observed in the two studies for different values of the kappa index imposed to the electron VDF at the base of the TR. For $\kappa \leq 4.5$ in [8] and $\kappa \leq 6$ in [3], the normalized heat flux is positive, that is in the direction of increasing temperature.

Finally both studies conclude that the heat conduction can be properly described by the classical Spitzer-Härm law only if the kappa index is sufficiently large, that is, if the VDF imposed at the base of the TR is close to a Maxwellian. As we said previously, this condition does not seem to occur according to [34], who claim that distributions with high-energy tails may develop in the TR.

3.3 Exospheric Models of the Solar Wind

Let us examine now what would be the consequences for the solar wind acceleration if non-thermal electron VDFs were present at the base of the wind. This problem has been extensively studied in the frame of the so-called Solar

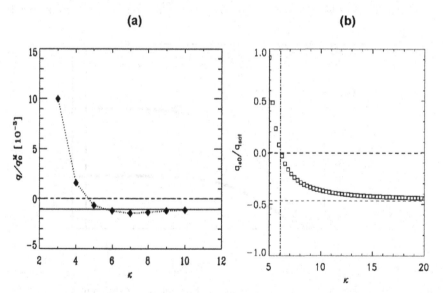

Fig. 6. The total normalized heat flux observed in the two studies by [8] (**a**) and [3] (**b**), for different values of the kappa index imposed to the electron VDF at the base of the transition region

Wind "exospheric" models [9, 12, 13, 15, 40, 41]. In these models, the plasma is assumed to be completely collisionless beyond a given altitude, called the exobase. In principle, the collisionless nature of the plasma above the exobase allows the computing, for each particle species, of the VDF at any arbitrary height as a function of the VDF at the exobase, by means of Liouville's theorem. However, the task is not trivial because the electric field profile needed to ensure local quasi-neutrality and zero current is an unknown of the problem that has to be determined self-consistently. The electric field arises because of the small electron-to-proton mass ratio that makes it easier for an electron, compared to a proton of the same energy, to escape from the star. In short, the electric force must be directed toward the sun for the electrons and away from the sun for the protons. This field is thus responsible for the strong outward acceleration of the protons [15].

Recently [12] and [40] proposed the most complete exospheric model of the solar wind, with a transonic solution for the speed profile. As an illustration of the results of the model by [40], we reproduce in Fig. 7 the radial evolution of the solar wind bulk speed obtained with Kappa VDFs as boundary conditions for the electrons at the exobase. A high terminal bulk speed (larger than 700 km/s) is obtained when the electron suprathermal tail is conspicuous ($\kappa = 2.5$). This is due to the interplanetary ambipolar electric field, which, due to the necessity to ensure the local quasi-neutrality and zero current, increases with the increasing number of suprathermal electrons at the base of the wind. As a consequence the ambipolar electric field accelerates the solar wind more than in the case when distributions are closer to a Maxwellian ($\kappa = 6$). An important remark to note in Fig. 7 is that the major part of this high terminal bulk speed

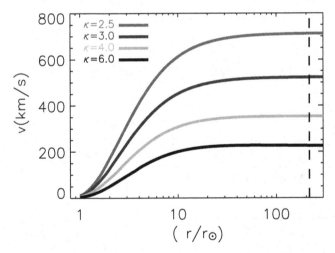

Fig. 7. Bulk speed profiles in the Zouganelis et al. model for different values of the parameter *kappa* representing the strength of the suprathermal tail ($\kappa = 2.5$, 3, 4, and 6. The *dashed vertical line* indicates the Earth orbit

is obtained within a small heliocentric distance ($\approx 10R_s$). This is due to the large acceleration represented by the large slope of the total (gravitational plus electric) ion potential in this region. Finally [40] have demonstrated that the high value of the terminal bulk speed, obtained in exospheric models, is not just an artefact of the use of Kappa functions. Such speeds can be obtained also with a sum of two Maxwellians (a cold and a hot one), which is a most classical model to represent the observed electron distributions.

4 Concluding Remarks

The solar wind is a highly collisionless medium where high-energy tails are always present. We have seen that this property is also true for the corona and even the temperature transition region, where Knudsen numbers as low as 10^{-3} to 10^{-2} invalidate the classical Spitzer-Härm theory. Non-Maxwellian electron velocity distributions could therefore also develop in the corona and an increasing amount of both theoretical and observational works are investigating this possibility.

Recent kinetic models have demonstrated that if such distributions exist in the high chromosphere, they can survive collisions through the transition region and "heat" the corona in a way which is decoupled from the local temperature gradient. Although the presence of non-thermal distributions in the chromosphere is still being discussed, mechanisms describing their formation in such a collisional medium have already been proposed [36].

Then finally if non-thermal electron distributions exist in the corona, they could participate to the wind acceleration, through the ambipolar electric field which is needed to ensure the local quasi-neutrality and zero current.

However, there is still a lot to do in the area covered in this article and major questions remain. First of all if non-thermal distributions exist in the high chromosphere, how are they formed there? Also the question whether the non-thermal character of the electron distributions in the solar wind is a consequence of transport and diffusion effects or whether it is also the cause of the acceleration is still under debate [19]. The answers to these questions may have to wait a space probe to visit the solar wind acceleration region of the corona.

References

1. C. Chiuderi and F. Chiuderi Drago: Astron. Astrophys., **422**, 331 (2004)
2. H.G. Demars and R.W. Schunk: Plan. Space Sci., **38**, 1091 (1990)
3. J.C. Dorelli and J.D. Scudder: J. Geophys. Res., **108**, 1294 (2003)
4. R. Esser and R.J. Edgar: Astrophys. J., **532**, L71 (2000)
5. W.C. Feldman, J.R. Asbridge, S.J. Bame et al, J. Geophys. Res., **80**, 4181 (1975)

6. R. Fisher and M. Guhathakurta: Astrophys. J., **447**, L139 (1995)
7. K. Issautier, N. Meyer-Vernet, M. Moncuquet et al: J. Geophys. Res., **103**, 1969 (1998)
8. S. Landi and F.G.E. Pantellini: Astron. Astrophys., **372**, 686 (2001)
9. K. Jockers: Astron. Astrophys., **6**, 219 (1970)
10. Y.K. Ko, J. Geiss and G. Gloeckler: J. Geophys. Res., **103**, 14539 (1998)
11. S. Koutchmy: Solar Physics, **51**, 399 (1977)
12. H. Lamy, V. Pierrard, M. Maksimovic et al, J. Geophys. Res., **108**, 1047 (2003)
13. J. Lemaire and M. Scherer: J. Geophys. Res., **76**, 7479 (1971)
14. M.P. Leubner: Astrophys. Space Sci., **282**, 573 (2002)
15. M. Maksimovic, V. Pierrard and J. Lemaire: Astron. Astrophys., **324**, 725 (1997)
16. M. Maksimovic, V. Pierrard and P. Riley: Geophys. Res. Letters, **24**, 1151 (1997)
17. M. Maksimovic, S.P. Gary and R.M. Skoug: J. Geophys. Res., **105**, 18337 (2000)
18. M. Maksimovic, V. Pierrard and J. Lemaire: Astrophys. Space Sci., **277**, 181 (2001)
19. M. Maksimovic, I. Zouganelis, J.-Y. Chaufray et al: J. Geophys. Res., **110**, A09104, doi:10.1029/2005JA011119 (2005)
20. E. Marsch, K.-H. Mühlhäuser, R. Schwenn et al: J. Geophys. Res., **87**, 52 (1982)
21. J.C. Maxwell: Nature, **8**, 537 (1873)
22. N. Meyer-Vernet, M. Moncuquet and S. Hoang: Icarus, **116**, 202 (1995)
23. N. Meyer-Vernet, A. Mangeney, M. Maksimovic et al: *Solar Wind Ten: Proceedings of the Tenth International Solar Wind Conference*, American Institute of Physics Conf. Proc., **679**, 263 (2003)
24. M.D. Montgomery, S.J. Bame and A.J. Hundhausen: J. Geophys. Res., **73**, 4999 (1968)
25. A. Pannekoek: Bull. Astron Inst. Neth., **1**, 107 (1922)
26. E.N. Parker: *Interplanetary dynamical processes*, (Interscience Pub., New York 1963)
27. W.G. Pilipp, H. Miggenrieder, M.D. Montgomery et al: J. Geophys. Res., **92**, 1075 (1987)
28. D.J. Pinfield, F.P. Keenan, M. Mathioudakis et al: Astrophys. J., **527**, 1000 (1999)
29. H. Rosenbauer, R. Schwenn, E. Marsch et al: J. Geophys., **42**, 561 (1977)
30. M.N. Rosenbluth, W.M. MacDonald and D.L. Judd: Phys. Rev., **107**, 1 (1957)
31. S. Rosseland: Mon. Notic. Roy. Astron. Soc., **84**, 720 (1924)
32. J.D. Scudder: Astrophys. J., **398**, 299 (1992)
33. J.D. Scudder: Astrophys. J., **398**, 319 (1992)
34. E.C. Shoub: Astrophys. J., **266**, 339 (1983)
35. L. Spitzer and R. Härm: Phys. Rev. A, **89**, 372 (1953)
36. A.F. Viñas, H.K. Wong and A.J. Klimas: Astrophys. J., **528**, 509 (2000)
37. C. Vocks and G. Mann: Astrophys. J., **593**, 1134 (2003)
38. K. Wilhelm, E. Marsch, B.N. Dwivedi et al: Astrophys. J. **500**, 1023 (1998)
39. G.L. Withbroe: Astrophys. J., **325**, 442 (1988)
40. I. Zouganelis, M. Maksimovic, N. Meyer-Vernet et al: Astrophys. J., **606**, 542 (2004)
41. I. Zouganelis, N. Meyer-Vernet, S. Landi et al: Astrophys. J., **626**, L117, (2005)

Recent Research: Large-scale Disturbances, their Origin and Consequences

Gottfried Mann[1] and Bojan Vršnak[2]

[1] Astrophysikalisches Institut Potsdam, An der Sternwarte 16, D-14482 Potsdam, Germany
gmann@aip.de
[2] Hvar Observatory, Faculty of Geodesy, Kačićeva 26, HR-10000 Zagreb, Croatia
bvrsnak@geof.hr

Abstract. This article gives a flavour of recent research dedicated to the large-scale coronal disturbances and the related interplanetary phenomena. The discussions include the take-off and propagation of coronal mass ejections (CMEs); the CME-flare relationship; the origin and propagation of shocks; the role of flares, CMEs, and shocks in particle acceleration; radio signatures of CMEs and shocks; coronal and IP plasma diagnostics offered by the radio emission excited by these phenomena.

1 Introduction

Large-scale disturbances, e.g., coronal mass ejections (CMEs), global waves (EIT and Moreton waves), shock waves, beams of energetic particles, global changes of the magnetic field topology and the related phenomena, dramatically affect the corona and the interplanetary (IP) space. Consequently, these phenomena directly or indirectly govern the solar-terrestrial relationship. Recently, the huge solar events that occurred in October/November 2003, have shown the vital public interest in solar activity: such events influence the Earth's environment and thus also our socio-economic activities. Since all these processes are rooted in the solar corona, studies of the coronal large scale disturbances lead us to a better understanding of Space Weather.

This report is based on discussions at the CESRA workshop at Sabhal Mor Ostaig, addressing the take-off and propagation of CMEs; the CME-flare relationship; the origin and propagation of shocks; the role of flares, CMEs, and shocks in particle acceleration; radio signatures of CMEs and shocks; coronal and IP plasma diagnostics offered by the radio emission excited by these phenomena.

G. Mann and B. Vršnak: *Recent Research: Large-scale Disturbances, their Origin and Consequences*, Lect. Notes Phys. **725**, 203–218 (2007)
DOI 10.1007/978-3-540-71570-2_10

2 Topics

The working group discussions addressed directly or indirectly the following specific topics:

- Launch and propagation of CMEs:
  - pre-eruption environment, magnetic structure, and processes;
  - basic kinematics and forces driving the eruption;
  - the role of reconnection;
  - radio signatures of the lift-off;
  - aerodynamic drag and forehead shock formation;
  - formation of current sheets (below and ahead/aside);
  - magnetic field restructuring and coronal dimming;
  - interruption of electron beams by CMEs;
- Origin and propagation of coronal and IP shocks
  - basic characteristics and terminology;
  - where and how do we see, or expect to see shocks (traveling & standing);
  - origin of coronal shocks: flares and/or CMEs;
  - formation and propagation of shocks in a decreasing/increasing Alfvén-speed and density environment;
  - multi-wavelength observations of Moreton waves;
  - nature of type II burst emission;
  - 3-dimensional propagation of MHD shocks in the magnetically structured atmosphere;
  - radio and optical signatures of coronal simple-waves and shocks;
- Particle acceleration in flares, CMEs, and shocks
  - acceleration at the CME forehead shock;
  - acceleration and escape from the current sheet below the CME (two-ribbon flare);
  - electron acceleration at the reconnection outflow (termination) shock
  - trapping and escape processes;
- Coronal & IP plasma diagnostics
  - m-km type II burst revealing physical conditions in the corona and IP space;
  - m-km type II bursts as CME tracers;
  - type III bursts mapping the magnetic field;
  - Moreon/EIT-waves as a diagnostic tool for low corona;
  - probing the corona by the radar techniques.

3 Take-off and Propagation of CMEs

The launch of a CME is usually accompanied by various radio-emission signatures. The most distinct ones are certainly the meter-to-kilometer wavelength radio type III bursts [50], revealing the escape of electron beams from the corona to the IP space along open field lines. However, sometimes the electron beams in the CME take-off phase remain confined to the corona, e.g., when the electrons are channeled along the large-scale coronal loops, or when the beams are interrupted by the CME itself. Such processes might be important for the understanding of the magnetic configuration involved in the eruption.

Silja Pohjolainen presented observations of a series of radio J-bursts observed by the Nançay Radioheliograph in the early stage of a CME lift-off on 8 February 2000 (see [45]). Type J bursts are interpreted as the radio signature of electron beams that propagate from an acceleration region in one leg to the vicinity of the summit of the loop, where they fade, which gives a signature in the time-inverse frequency plane that resembles the inverted letter J. The observed radio pattern was interpreted as a radio signature of the break-out process similar to that proposed by [2]. In the follow-up discussion, the event was compared to the famous 2 May 1998 event [44], since transequatorial coronal loops were involved in both events, similar to those observed by [23].

The discussion turned to a debate regarding the driving process of the eruption. In the framework of the ongoing observational and theoretical research, Bojan Vršnak argued that the basic process which causes the eruption is loss of equilibrium of the magnetic structure (e.g., [17, 33]). Plasma flows in the solar convective zone drive evolution of the coronal magnetic field, increasing complexity of its topology by building the induced electric current system, i.e., storing the free energy into the magnetic structure. The loss of equilibrium takes place when the slowly evolving system cannot find a neighbouring equilibrium state and undergoes catastrophic development (see, e.g., [46]), resulting in the eruption of the unstable magnetoplasmatic structure into the interplanetary space.

Bearing in mind the geometry of the ejection, the dynamics of the process itself is most likely governed by the kink instability of the flux rope (e.g., [58], and references therein). Since coronal eruptions and their interplanetary counterparts often expose helical patterns (e.g., [28, 68]), it is assumed that the ejection is basically driven by the hoop force within the helically twisted semi-toroidal magnetic structure [3, 10, 11, 43, 63], i.e., within the flux rope nested in a coronal magnetic arcade [35]. The instability sets in when the magnetic-pressure gradient associated with the poloidal magnetic field component in the flux rope becomes so large that it cannot be balanced by the magnetic tension of the longitudinal magnetic field component.

According to this view, the processes like, e.g., break-out process [2], or tether cutting [42] are only modifying the course of the eruption. On the other hand, this type of processes could play an important role in the pre-eruptive stage and at the onset of the eruption, since the reconnection below (tether cutting) or above (break out) can drive the evolution of the system towards the critical point where the equilibrium is lost. Possible drivers are also the emerging flux process, twisting/shearing motions at the footpoints, the mass loss from the prominence, or their combination.

Furthermore, it was pointed out by B. Vršnak that the reconnection above the erupting flux rope is less important in 3-dimensional (3-D) models since in real situation the overlying field-lines can "slip aside" the erupting flux rope (see e.g., Fig. 2 of [14]). It was concluded that the 3-D modeling of the eruption of the flux rope anchored in the photosphere, like currently being performed by various research groups (e.g., [1, 14, 29, 51]), might be essential for the comprehension of the CME take-off.

Furthermore, it was emphasized that special attention should be paid to the role of the reconnection below the flux rope, causing the two-ribbon flare. This process provides a supply of the poloidal magnetic flux to the rope, enhancing the kink effect (Fig. 1). Since the reconnection is also causing the flare energy release, such a coupling of the CME dynamics and the reconnection could explain the synchronization of the CME acceleration and the growth of the associated soft X-ray burst (e.g., [74]; see also [66] and references therein). The hypothesis of the feed-back relationship between the CME acceleration and the reconnection rate in the associated flare is also consistent with the correlation between the CME speed and the flare importance, illustrated in Fig. 2 (for details see [69]).

In this respect, the presentation by Lyndsay Fletcher (see Sect. 5) was also partly related to the CME take-off, since she presented an ongoing research on the relationship between the energy release in the CME-associated flare and the "motion" of flare kernels. The reconnection rate $\mathbf{v} \times \mathbf{B}$, inferred by using the velocity of the flare-ribbon lateral expansion and the line-of-sight photospheric magnetic field (e.g., [70]), provides an estimate of the energy-loading into the CME (see also, e.g., [32]). Several recent studies clearly demonstrated the correlation of the CME acceleration and the $\mathbf{v} \times \mathbf{B}$ evolution (e.g., [47, 49], and references therein). In the discussion it was concluded that it is extremely important to distinguish the reconnection-rate associated kernel motions, from that which occurs as a consequence of spreading of the reconnection along the neutral line (see, e.g., [19]), or the reconnection in a strongly sheared field (e.g., [7]). In her report, L. Fletcher demonstrated also how a combination of

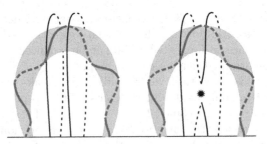

Fig. 1. Reconnection beneath the CME (*gray flux rope*) creates a two-ribbon flare (*low-lying loops in the right panel*) and supplies a "fresh" azimuthal flux to the rope (*the black helical line around the flux rope*)

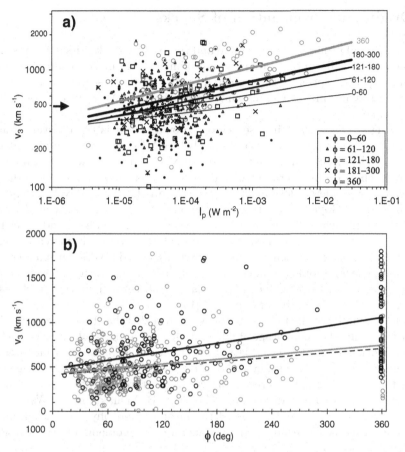

Fig. 2. a) Dependence of the CME velocities at the heliocentric distance of $R = 3$ solar radii (v_3) on SXR peak fluxes, I_p, of the associated flares (the relationship is shown separately for different classes of CME widths, denoted by the *fitted lines*). The average velocity $\bar{v}_3$ of non-flare CMEs associated with prominence eruptions is indicated by the *horizontal arrow* at the y-axis. **b)** The correlation of the CME velocity v_3 and the CME width ϕ: B & C-flare CMEs are shown in *gray* and M & X-flare CMEs are shown in *black*. The fit taken from the sample of non-flare CMEs is shown by the *dashed line*. From [69]

the multiwavelength flare observations and the magnetic field extrapolation can be used to reveal the magnetic topology involved in the eruption.

Finally, it was agreed that meticulous multi-wavelength studies of the ejection/flare morphology are needed to distinguish signatures of relevant processes governing the CME take-off. Such studies, offered by the existing excellent space-borne and ground based instruments, might be also very important for the Space Weather research, in particular for the CME arrival-time predictions.

4 Origin and Propagation of Shocks

The origin and propagation of coronal and IP shocks are a key issue in present research that will continue to play a prominent role in the coming years. Several aspects, especially those related to the ongoing research in this field, are discussed also in Nat Gopalswamy's and Alexander Warmuth's articles in this volume.

A long standing issue in research on coronal shocks is the question if all coronal waves/shocks are driven exclusively by CMEs, or if they can be ignited by flares. The discussion was introduced by N. Gopalswamy, who presented a progress report on the hierarchy of CME characteristics, with respect to their association with meter, dekameter/hectometer, and kilometer wavelength type II bursts ([18]; Gopalswamy, this volume). Generally, CMEs associated with type II bursts have larger than average speeds, angular widths, and decelerations, and the values of these three parameters successively increase from meter, to kilometer wavelength type II burst associated CMEs. Such a hierarchy, and some additional arguments (see Gopalswamy, this volume), were used as an evidence supporting the hypothesis that all coronal shocks are driven by CMEs (for historical background, cf. [13]).

The beginning of the follow-up discussion revealed that there is a great diversity in comprehension/usage of different physical terms, such as bow-shock, piston-shock, spherical piston, blast-wave, simple-wave, pressure or magnetic-pressure pulse, shock flanks, etc. Confusing situation regarding the usage of these terms is present also in recent, as well as older, related literature. So, before proceeding with the discussions, G. Mann, B. Vršnak, and A. Warmuth gave a brief overview, and discussed the meaning and physical background of these terms (see Warmuth, this volume; for the problems in terminology see [69]).

After clarifying the terminology, the discussion turned back to various aspects of the relationship between CMEs, flares, coronal dimming, type II bursts, Moreton waves, and EIT waves. A. Warmuth emphasized that sometimes a type II burst and/or a Moreton wave occurs at the time when the CME is already quite high in the corona, which contradicts the hypothesis according to which all coronal shocks are driven by CMEs. In such events the CME-associated shock is located at radial distances considerably larger than that of the observed type II burst sources (especially in the case of high-frequency type II bursts that start in the decimeter/meter wavelength range), whereas the flanks of the shock are too far to cause the Moreton wave. It was also pointed out that in the meter-type II burst sample presented by N. Gopalswamy there is a considerable fraction of slow CMEs which are not likely candidates for driving shocks.

Furthermore, B. Vršnak demonstrated that CMEs whose bow-shock should cause a Moreton wave at the distance in the order of 100 Mm, and a type II burst of the starting frequency around 300 MHz, would need to have extremely large acceleration, up to 10 km s^{-2}, whereas accelerations larger than 1 km s^{-2}

are observed very rarely (the highest measured acceleration was inferred in the event of 6 November 1997; see [13]). Moreover, the short time involved in the shock formation (a few minutes) requires an abrupt/fast source region expansion: the acceleration should not last more than a few minutes, which is a time-scale much more appropriate for flares than for CMEs. Additional evidence supporting the view that at least some coronal shocks are ignited by flares was noted by J. Khan, who emphasized the close timing and spatial association of type II bursts and Moreton/EIT/SXR waves with the impulsive energy release in flares [33, 72]. However, in the follow-up discussion, it was also pointed out that such an argument is weakened by the fact that the acceleration phase of CMEs is often tightly related to the impulsive phase of the associated flare ([74]; see also [66] and references therein).

The discussions on the origin of coronal shocks ended up with the conclusion that the next step in the research should be a combined analysis which would include not only meticulous study of multi-wavelength observations, but should also incorporate the theoretical knowledge on the shock formation phase (e.g., [37, 65]), characteristics of 3-D propagation (e.g., [73]), the relative kinematics of the shock and its driver [65], the shock–driver offset distance [52], etc.

5 The Role of Flares, CMEs, and Shocks in Particle Acceleration

A considerable part of the WG-II time-table was devoted to the particle acceleration associated with the eruptive processes in the solar corona. Several observational and theoretical aspects were discussed, including different possible accelerators and mechanisms of the escape of accelerated particles. In particular, the acceleration by the electric field involved in reconnection (the diffusion region and slow-mode shocks), the acceleration at CME driven shocks, and the acceleration at the fast-mode standing shock in the reconnection outflow jet were considered.

Astrid Veronig and Manuela Temmer presented an ongoing study of the 3 November 2003 X4 flare, emphasizing the initial downward motion of the loop-top hard X-ray (HXR) source, as well as the "motion" of the HXR foot-point sources and their positions relative to the underlying photospheric magnetic field. The downward motion of the loop-top HXR source is currently subject of intense research ([72]; see also [62], and references therein, and the chapters by Dennis et al. and Hudson & Vilmer, this volume). It was shown by A. Veronig and M. Temmer that the main energy release starts after the downward motion of the loop-top source stops and the "standard" growth of the hot loop-system starts. The discussion included various interpretations of the downward motion of the loop-top source. The interpretation according to which such a motion reveals the formation of the reconnection outflow jets prevailed. An ongoing research in this field includes the betatron acceleration in the collapsing trap formed by shrinking loops in the deflection sheath of the reconnection jet [30, 62].

The issue of the motion of foot-point sources was stressed most directly by Lyndsay Fletcher, who showed new results regarding the relationship between the product $\mathbf{v}_{kernel} \times \mathbf{B}_{ph}$ (the kernel velocity times the underlying photospheric field) and the energy release in the flare. It was shown that sometimes there is a direct correlation between the value of $\mathbf{v}_{kernel} \times \mathbf{B}_{ph}$ and the energy release rate, but situations without any correlation were also spotted. After a dynamical discussion, it was concluded that most probably, the very high values of $\mathbf{v}_{kernel} \times \mathbf{B}_{ph}$ that are sometimes measured, do not reflect the electric field involved in the reconnection, but are rather caused by a successive activation of new energy release sites along the neutral line (see, e.g., [19]), or the reconnection in a strongly sheared field (e.g., [7]). Such phases of the energy release do not necessarily lead to an enhanced energy release. Recently, a number of studies were performed, where various proxies for the reconnection rate were compared with the energy release in the flare (e.g., [15, 16, 48, 70]). Further discussions included the acceleration process itself. Gottfried Mann presented an application of the shock drift acceleration mechanism to the fast-mode standing shock in the reconnection outflow ("termination shock"; [4, 5]). It was demonstrated that a significant number of high-energy electrons can be produced at the termination shock by shock drift acceleration [6, 20, 40], sufficient to explain non-thermal energy release in flares. In contrast to that, Karl-Ludwig Klein deduced from two particular events (August 19, 1996 and October 7, 1997), that the number and energy "fluxes of shock-accelerated electrons are much smaller than those required to produce a conspicuous hard X-ray burst" [26]. But [26] considered travelling shocks as usual type II burst sources, i.e. they appear at low frequencies (i.e. < 100 MHz), in the higher corona. There, the electron number density is quite smaller than in regions, where the "termination shock" sources occur, i.e. around 300 MHz. Additionally, the temperature is much higher at the "termination shock" (i.e. > 10 MK), i.e. near the reconnection site, than in the non-flaring corona (≈ 1.4 MK). Both are the reasons, why the "termination shock" is much more productive for the generation of energetic electrons, than the travelling shocks, which are the sources of the usual type II bursts. Furthermore, an example of the radio signature of the termination shock was presented: a non-drifting type II-like burst was observed in 18 July 2002 flare, in conjunction with a distinct impulsive hard X-ray burst.

Observations of this kind, especially when combined with measurements of the flare loops and the collapsing trap theory, could give a better insight into the multi-stage acceleration process in flares. This includes the direct-field-acceleration in the diffusion region (e.g., [34]) and at the slow mode shocks [54], shock-drift acceleration at the termination shock [5, 56, 60], and betatron acceleration in the deflection sheath beneath the termination shock [30]. All of these processes should show different particle-spectral characteristics (thermal/nonthermal), which could differ in different phases of the energy release, as well as in flares taking place in different environments. Moreover, each process should show distinct radio signatures, depending on details of particle phase-space distribution (bump-on-tail, loss-cone, etc.).

Another aspect of the particle acceleration was discussed by Karl-Ludwig Klein, who presented an analysis [41] of the arrival of high-energy particles at 1 AU. Fast CMEs without flares were checked to inspect the CME-accelerated particles in the absence of flares. In the period 1996-1998 only three such events were found. In two of these, no SEP[1] fluxes at 1 AU were detected, and in the remaining one, only a weak enhancement limited to energies $\leq 20\,\mathrm{MeV}$ was recorded. That indicates that the energy release in flares is important for the emission of high-energy particles into IP space [12, 25, 26, 27]. In the follow-up discussion several escape mechanisms were considered, including direct involvement of open field lines in the energy release process and the temporary opening of the field lines by reconnection with nearby coronal hole field. A supporting evidence for the flare-accelerated IP electron beams was demonstrated by Laura Bone, who presented an example of the hard X-ray burst, correlated with a type III burst which extended to long wavelengths, showing that electrons accelerated in solar flares can escape to the IP space.

Niina Lehtinen presented very interesting decimetric emission recorded in the event of 18 December 2000 (see [31]). Observations could be interpreted as the radio emission from the rising plasmoid [30]. Furthermore, the trajectories of electron beams were imaged in radio-waves, from the acceleration site to the corresponding hard X-ray sources at the ends of soft X-ray loops. This event included a number of issues that are difficult to interpret. For example, the radio precursor and the associated electron accelerator were found to be very far away from the active region, located in between the EIT wave and the active region. In the follow-up discussion it was concluded that in this event, like in many others, there is a clear departure from the classical two-ribbon flare model.

Moreton and EIT waves are often associated with SEP events [8, 30, 59, 61]. There are some observational hints that the SEP had been released at times when the EIT wave touches the magnetic field line connecting the acceleration site, i.e. the EIT wave, in the corona with the spacecraft. As well-known, the EIT wave represents a "simple fast magnetosonic wave" [37, 38, 65] accompanied with a magnetic field compression. Thus, a EIT wave can act as a moving magnetic mirror, at which particles can be accelerated, e.g. by shock drift acceleration [6, 20, 40].

---

[1] Solar Energetic Particles.

6 Coronal and IP Plasma Diagnostics Offered by the Radio Emission

Solar eruptive phenomena are associated with a wide palette of radio emissions which offer various plasma diagnostic techniques, enabling an insight into physical conditions in the corona and IP space. At WG-II, special attention was paid to the Alfvén speed, v_A, in the corona and IP space. Bojan Vršnak presented results on the radial dependence of the Alfvén speed, $v_A(R)$, and the magnetosonic speed $v_{ms}(R)$ based on the measurements of the type II bursts. The method utilizes the old idea by [55], according to which the type II burst band-split is a consequence of the plasma emission from the upstream and downstream shock regions (see e.g., [40]). The results presented by B. Vršnak include the range from the decimeter/meter wavelengths up to dekameter/kilometer wavelengths. The results show (Fig. 3) that above active regions the magnetosonic velocity first decreases from 1000-2000 km s^{-1} to a minimum of ≈ 400 km s^{-1} at the radial distance $R \approx 3$ solar radii, then increases to a local maximum of ≈ 500 km s^{-1} at $R \approx$ 4–6 solar radii, and finally monotonously decreases in the IP space down to ≈ 30–50 km s^{-1} at 1 AU. As a side result, the radial dependence of the density was obtained, smoothly joining the active region corona with the IP space.

Gottfried Mann presented results of a related study, where the coronal Alfvén and magnetosonic speeds were modeled using a three-component input: the coronal density model, the global/radial magnetic field, and the bipolar

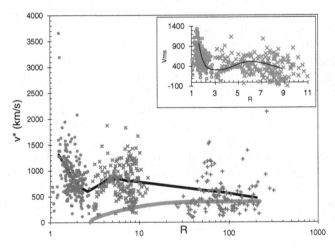

Fig. 3. Shock velocities v^* inferred from the frequency drift of type II bursts (after [67]), compared with the solar wind speed ($w(R)$; shown by *gray line*) according to [53]. In the inset we show the estimated magnetosonic speed $v_{ms}(R)$ in the corona and upper corona (negative values are the measurement artifact appearing when $v^* < w$)

active region field [38, 39, 72]. The model provides radial and angular dependence of the Alfvén speed (Fig. 4), showing a pattern consistent with the type II burst observations presented by B. Vršnak (compare Figs. 3 and 5). The bright and dark areas present the regions of high and low Alfvén speeds, respectively. A region of a local minimum of the Alfvén speed is located in the middle of the corona (i.e. in the lower half of the grey scaled area) (Fig. 4).

The follow-up discussion was focused on the implications regarding coronal shock propagation. The angular dependence of v_A shows that the low Alfvén velocity regions are located aside of active region which explains the tendency of non-radial motion of the radio source in the early stage of type II bursts. The distance where the Alfvén speed attains minimum, corresponds to plasma frequencies at which most of metric type II bursts start, whereas the maximum corresponds to the frequency range where most of the meter/dekameter

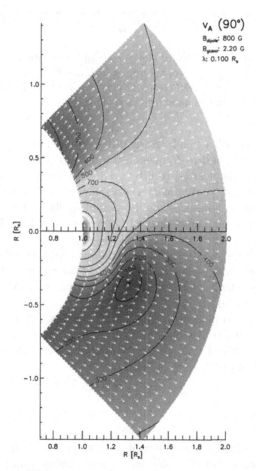

Fig. 4. Behaviour of Alfvén velocity given in km/s above a bipolar active region in the corona (see [72]). The arrows indicate the direction of the magnetic field

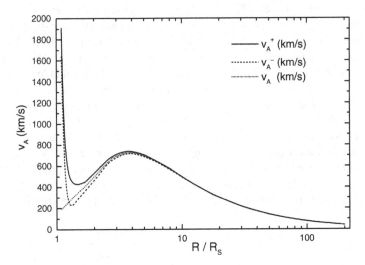

Fig. 5. Behaviour of Alfvén velocity along a straight line with an inclination of $45°$ towards the z-axis as a function of the radial distance R. The full and *dashed line* show the case of $\phi = 0°$ $\phi = 180°$, respectively. The *dotted line* represents the case only due to the quiet Sun magnetic field (see [39])

bursts cease. On the other hand, the distances beyond which $v_A(R)$ decreases monotonously, correspond to frequencies typical for the onset of IP type II bursts (see, e.g., [9]) that are driven by CMEs. Apparently, such an outcome supports the hypothesis that at least a certain fraction of metric type II bursts are a distinct phenomenon, presumably ignited by flares. On the other hand, the IP type II bursts occur after the CME of a given speed reaches the distance at which v_A becomes lower than the CME speed.

Another interesting aspect of the coronal plasma diagnostic, was stressed by M. Khotyaintsev, who presented a theoretical background of the solar radar experiments. In particular, a theory for radar signal scattering by anisotropic Langmuir turbulence in the solar corona due to a $t + l = t$ process was considered. The method is supposed to enable investigation of narrow field-aligned electron beams, such as those responsible for type III solar bursts. Expressions for the scattering altitudes, frequency shifts, cross-sections, efficiency of the process, and optical depth were presented.

7 Conclusion

The WG-II presentations initiated a plenty of dynamical discussions on several important aspects of the solar eruptive phenomena and large scale disturbances. From our point of view there are open questions, which are listed below without expecting any completeness:

- What causes the launch of a CME?
- What is the relationship between the flare and the initiation of the CME launch?
- What is the source region of a type II radio burst with respect to the shock?
- What is the mechanism of electron acceleration at a coronal shock wave?
- Is a coronal shock wave able to accelerate protons also seen at 1 AU?
- What is the reason of the delay of the arrival time of relativistic electrons at 1 AU concerning the onset of the energy release in the corona?
- What is the role of global travelling EIT waves for particle acceleration?
- In which way are these EIT waves related to the flare and the launch of a CME?

Many of the discussions continued also after the meeting, by means of electronic communications, including also colleagues who did not attend the Isle of Skye meeting. Some of the ideas were spread through the solar community at other meetings, and resulted in a number of ongoing studies. Some results of collaborations initiated at the CESRA/WG-II already have been presented in a number of scientific papers. In the near future we will have some space missions as STEREO, Hinode and Solar Dynamics Observatory (SDO). We expect that the data of these missions will give us a deeper insight in the physical processes in the solar corona and the related studies will hopefully give us the answers of a few of the questions mentioned above. Radio observations are an important addition because they are unique tracers of energy release and shock waves in the solar corona.

Acknowledgements

We wish to thank the organizers of the CESRA workshop for a particularly stimulating meeting, and the staff of Sabhal Mor Ostaig for their kind hospitality. BV acknowledges the financial support by CESRA organizers which enabled him to attend the meeting. The report was prepared during the stay of BV at AIP, which was financed by DFG 436 KRO 113/8/0-1.

Appendix: The List of Participants

The WG-II was attended by the following participants:
Laura Bone (laura@astro.gla.ac.uk)
Lyndsay Fletcher (lyndsay@astro.gla.ac.uk)
Nat Gopalswamy (gopals@fugee.gsfc.nasa.gov)
Josef Khan (jik@astro.gla.ac.uk)
Mykola Khotyaintsev (ko@irfu.se)
Karl-Ludwig Klein (ludwig.klein@obspm.fr)

Niina Lehtinen (nijole@utu.fi)
Gottfried Mann (Gmann@aip.de)
Silja Pohjolainen (silpoh@utu.fi)
Ralf Schroeder (frs@tp4.rub.de)
Manuela Temmer (mat@igam.uni-graz.at)
Astrid Veronig (asv@igam.uni-graz.at)
Bojan Vršnak (bvrsnak@geodet.geof.hr)
Mikko Kalervo Vaananen (mikko.vaananen@astro.helsinki.fi)
Alexander Warmuth (awarmuth@aip.de).

At several occasions a number of other participants of the workshop joined the WG-II discussions, mostly related to the particle acceleration.

References

1. Amari, T., Luciani, J. F., Mikic, Z., & Linker, J.: Astrophys. J. **529**, L49 (2000)
2. Antiochos, S. K., de Vore, C. R., & Klimchuk, J.: Astrophys. J. **510**, 485 (1999)
3. Anzer, U.: Solar Phys. **57**, 111 (1978)
4. Aurass, H., Vršnak, B., & Mann, G.: Astron. Asstrophys. **384**, 273 (2002)
5. Aurass, H. & Mann, G.: Astron. Asstrophys. **615**, 526 (2004)
6. Ball, L. & Melrose, D. B.: Publ. Astron. Sec. Austr. **18**, 361 (2001)
7. Bogachev, S. A., Somov, B. V., Kosugi, T. & Sakao, T.: Astrophys. J. **630**, 561 (2005)
8. Bothmer, V., Posner, A., Kunow, H., & 14 co-authors: Proc. 31st ESLAB Symposium **ESA-SP-415**, 207 (1997)
9. Cane, H. V. & Erickson, W. C.: Astrophys. J. **623**, 118 (2005)
10. Chen, J.: Astrophys. J. **338**, 453 (1989)
11. Chen, J. & Krall, J.: J. Geophys. Res. **108** DOI10.1029/2003JA009849 (2003)
12. Classen, H. T., Mann, G., Klassen, A., & Aurass, H.: Astron Astrophys. **409** 309 (2003)
13. Cliver, E. W., Nitta, N. V., Thompson, B. J., & Zhang, J.: Solar Phys. **225**, 105 (2004)
14. Fan, Y.: Astrophys. J. **630**, 543 (2005)
15. Fletcher, L. & Hudson, H. S.: Solar Phys. **210**, 307 (2002)
16. Fletcher, L., Pollok, J. A., & Potts, H. E.: Solar Phys. **222**, 279, (2004)
17. Forbes, T. G.: J. Geophys. Res. **95**, 11911 (1990)
18. Gopalswamy, N., Aguilar-Rodriguez, E., Yashiro, S., Nunes, S., Kaiser, M. L., Howard, R. A.: J. Geophys. Res. **110**, A12S07, 10.1029/2005JA011158 (2005)
19. Grigis, P. C. Benz, A. O.: Astrophys. J. **625**, L143 (2005)
20. Holman, G. D. & Pesses, M. E.: Astrophys. J. **267**, 837 (1983)
21. Karlický, M.: Astron. Astrophys. **414**, 325 (2004)
22. Karlický, M. & Kosugi, T.: Astron. Asstrophys. **419**, 1159 (2004)
23. Khan, J. I., Hudson, H. S.: Geophys. Res. Lett. **27**, 1083 (2000)
24. Khan, J. I. & Aurass, H.: Astron. Astrophys. **343**, 1018 (2002)
25. Klein, K.-L. & Posner, A.: Astron. Astrophys. **438**, 1029 (2005)
26. Klein, K.-L., Schwartz, R. A., McTiernan, J. M., Trottet, G., Klassen, A., & Lecacheux, A.: Astron. Astrophys. **409**, 317 (2003)

27. Klein, K.-L., Krucker, S., Trottet, G., & Hoang, S.: Astron. Astrophys. **431**, 1047 (2005)
28. Klein, L. W. & Burlaga, L. F.: J. Geophys. Res. **87**, 613 (1982)
29. Kliem, B., Titov, V. S., & Török, T.: Astron. Asstrophys. **413**, L23 (2004)
30. Krucker, S., Larson, D. E., Lin, R. P., & Thompson, B. J.: Astrophys. J. **519**, 864 (1999)
31. Lehtinen, N. J., Pohjolainen, S., Karlický, M., Aurass, H., Otruba, W.: Astron. Astrophys. **442**, 1049 (2005)
32. Lin, J., Raymond, J. C., & van Ballegooijen, A. A.: Astrophys. J. **602**, 422 (2004)
33. Lin, J., Soon, W., & Baliunas, S. L.: New Astron. Rev. **47**, 53 (2003)
34. Litvinenko, Yu.E.: Solar Phys. **212**, 379 (2003)
35. Low, B. C.: Solar Phys. **167**, 217 (1996)
36. Magdalenić, J., Vršnak, B., & Aurass H.: in *Proc. 10th European Solar Phys. Meeting: 'Solar Variability: From Core to Outer Frontiers'*, ESA SP-**506**, 335 (2002)
37. Mann, G.: J. Plasma Phys. **53**, 109 (1995)
38. Mann, G., Klassen, A., Aurass, H., Estel., C., & Thompson, B.: Proc. 8th SOHO Workshop **ESA SP-446**, 477 (1999)
39. Mann, G., Klassen, A., Aurass, H.,& Classen, H. T.: Astron. Astrophys. **400**, 614 (2003)
40. Mann G. & Klassen, A.: Astron. Astrophys. **441**, 319 (2005)
41. Marqué, C., Posner, A., & Klein, K.-L.: Astrophys. J. **642**, 1222 (2006)
42. Moore, R. L., LaRosa, T. N., Orwig, L. E.: Astrophys. J. **438**, 985, (1995)
43. Mouschovias, T. Ch. & Poland, A. I.: Astrophys. J. **220**, 675 (1978)
44. Pohjolainen, S., Maia, D., Pick, M., et al.: Astrophys. J. **556**, 421 (2001)
45. Pohjolainen, S., Vilmer, N., Khan, J. I., Hillaris, A. E.: Astron. Astrophys. **434**, 329 (2005)
46. Poston, T. & Stewart, I.: Catastrophe Theory and Its Applications, Pitman, San Francisco (1978)
47. Qiu, J., & Yurchyshyn, V. B.: Astrophys. J. **634**, L121 (2005)
48. Qiu, J., Lee, J., Gary, D. E., Wang, H.: Astrophys. J. **565**, 1335 (2002)
49. Qiu, J., Wang, H., Cheng, C. Z., & Gary, D. E.: Astrophys. J. **604**, 900 (2004)
50. Reiner, M. J., Kaiser, M. L. & Bougeret, J.-L.: J. Geophys. Res. **106**, 29989 (2001)
51. Roussev, I. I., Forbes, T., Gombosi, T. I., et al.: Astrophys. J. **588**, L45 (2003)
52. Russell, C. T. & Mulligan, T.: Planetary and Space. Sci. **50**, 527 (2002)
53. Sheeley, N. R., Wang, Y.-M., Hawley, S. H., et al.: ApJ, **484**, 472–478 (1997)
54. Shimada, N., Terasawa, T., & Jokipii, J. R.: J. Geophys. Res., **102**, 22301 (1997)
55. Smerd, S. F., Sheridan, K. V., & Stewart, R. T.: in *IAU Symp. 57*, ed. G. A. Newkirk, p. 389 (1974)
56. Somov, B. V. & Kosugi, T.: Astrophys. J. **485**, 859 (1997)
57. Sui, L., & Holman, G. D.: Astrophys. J. **596**, L251 (2003)
58. Török, T. & Kliem, B.: Astrophys. J. **630**, L97 (2005)
59. Torsti, J., Anttila, A., Kocharov, L., & 7 co-authors: Geophys. Res. Lett. **25**, 2525 (1998)
60. Tsuneta, S. & Naito, T.: Astrophys. J. **495**, L67 (1998)
61. Vainio, R. & Khan, J. I.: Astrophys. J. **600**, 451 (2004)

62. Veronig, A., Karlický, M., Vršnak, B., et al.: Astron. Astrophys. **446**, 675 (2006)
63. Vršnak, B.: Solar Phys. **129**, 295 (1990)
64. Vršnak, B.: EOS **86**, No.11, 112 (2005)
65. Vršnak, B. & Lulić, S.: Solar Phys. **196**, 157 (2000)
66. Vršnak, B., Maričić, D., Stanger, A., Veronig, A.: Solar Phys. **225**, 355 (2004a)
67. Vršnak, B., Magdalenic, J., & Zlobec, P.: Astron. Astrophys. **413**, 753–763 (2004b)
68. Vršnak, B., Ruždjak, V., & Rompolt, B.: Solar Phys. **136**, 151 (1991)
69. Vršnak, B., Sudar, D., & Ruždjak, D.: Astron. Astrophys. **435**, 1149 (2005)
70. Wang, H., Qiu, J., Ju, J., & Zhang, H.: Astrophys. J. **593**, 564 (2003)
71. Warmuth, A. & Mann, G.: Astron. Astrophys. **435**, 1123 (2005)
72. Warmuth, A., Mann, G., & Aurass, H.: Astrophys. J. **626**, L121 (2005)
73. Wu, S. T., Li., B., Wang, S., & Zheng, H.: J. Geophys. Res. **110**, A11102, 10.1029/2005JA011056 (2005)
74. Zhang, J., Dere, K. P., Howard, R. A., Vourlidas, A.: Astrophys. J. **604**, 420 (2004)

Plasma of the Solar Corona

Quasi-periodic Pulsations as a Diagnostic Tool for Coronal Plasma Parameters

V. M. Nakariakov[1] and A. V. Stepanov[2]

[1] Physics Department, University of Warwick, Coventry CV4 7AL, UK
valery@astro.warwick.ac.uk
[2] Pulkovo Observatory, Russian Academy of Sciences, Pulkovskoe sh. 65,
St Petersburg, 196140 Russia
stepanov@gao.spb.ru

Abstract. Quasi-periodic pulsations of radio emission generated in solar and stellar flares are interpreted in terms of MHD oscillations of coronal loops and in the frame of equivalent electric circuit approach, and compared with coronal wave and oscillatory phenomena recently discovered in the EUV, X-ray and visible light bands. Various methods of remote diagnostics of the coronal plasma, based upon the use of observationally detectable properties of quasi-periodic pulsations - their period, amplitude and quality - are discussed. The applicability of these methods to the diagnostics of stellar coronae is demonstrated.

1 Introduction

Wave and oscillatory phenomena in solar and stellar coronae have been attracting researcher's attention for several decades. Traditionally, the interest is motivated by the possible role the waves play in heating of the coronae and in acceleration of solar and stellar winds, via the transfer and deposition of energy and mechanical momentum. Also, as waves and oscillations are associated with various dynamical processes in the almost fully ionised and highly magnetised coronal plasma, their study is fundamental for plasma astrophysics in general. Moreover, the progress recently reached in observational detection and theoretical modelling of coronal waves and oscillations provided the foundation for the development of *coronal seismology* - a new and rapidly developing branch of astrophysics, which combines theoretical and observational findings in deducing information about physical parameters in the corona.

An important role in the interpretation of coronal wave and oscillatory phenomena is played by magnetohydrodynamic (MHD) theory. Indeed, the observed characteristic scales, e.g. wave lengths are usually much larger than both the ion gyroradius and the mean free path length and the characteristic times are much greater than the ion gyroperiod and the reciprocal of

V. M. Nakariakov and A. V. Stepanov: *Quasi-periodic Pulsations as a Diagnostic Tool for Coronal Plasma Parameters*, Lect. Notes Phys. **725**, 221–250 (2007)
DOI 10.1007/978-3-540-71570-2_11

the collision frequency. Compressibility and elasticity of coronal plasmas allow them to support various kinds of MHD waves which are associated with perturbations of macro parameters of plasmas: its mass density, temperature, gas pressure, bulk velocity, and the local direction and absolute value of the magnetic field. Detection of these perturbations in the coronal plasma as fluctuations of emission intensity of coronal lines, their Doppler shifts and non-thermal broadening, and radio emission provides us with the basis for the observational study of MHD waves in coronae.

The proximity of the solar corona and its openness to high spatial and temporal resolution observations make it an ideal object for the study of its wave activity. Here we would like to stress that normally the proper study of a wave requires both its projected wave length and period being well resolved. Only in certain exceptional cases, e.g. when the wave passes through a bright object (e.g., a part of a flaring loop) with the geometrical size smaller than the wavelength, the wave can be detected with poorer spatial resolution or even without it at all.

The necessary spatial and time resolution has recently been achieved in the EUV band with spaceborne imagers (TRACE, SOHO/EIT) and spectral instruments (SOHO/SUMER), leading to the avalanche of observational discoveries of solar coronal waves and oscillations. The discovered phenomena can be divided into several distinct classes: EIT waves [42]; compressible waves in polar plumes [11, 32, 33] and in coronal loops [8, 12]; flare-generated global kink oscillations of loops [5, 26]; and longitudinal standing oscillations within loops [19, 44]. Moreover, very recently ground-based optical observations of eclipses have revealed the presence of rapid oscillations in coronal loops [47, 48]. Extensive observational reviews of this diversity of coronal oscillations are given in [4, 24].

Ground-based radio observations also demonstrate various kinds of oscillations (e.g., the quasi-periodic pulsations, or QPP, see [2] for a review and Nindos & Aurass, this volume), usually with periods from a few seconds to several tens of seconds, often in association with a flare. However, usually those observations do not provide the spatial information and the only parameter which can be used for the interpretation of the pulsations is the period. According to the period P, Aschwanden [3] suggested to distinguish between *very short* period ($P < 0.5$ s), *short* period ($0.5 < P < 5.0$ s) and *long-period* ($P > 5$ s) coronal radio QPP. It is believed that QPP with periods longer than a second might be associated with MHD waves and oscillations. In particular, QPP with periods of about a second are commonly detected at metric, decimetric and centimetric wave lengths, usually in flaring radioflux curves, often in association with type IV bursts. A list of more than 30 papers on observations of periodicities in this range is presented in [3].

Periodic variations of physical quantities can be connected with several mechanisms, namely the presence of certain resonances connected with standing waves, the generation of quasi-periodic frequency- and amplitude-modulated signals because of dispersion, and nonlinear mechanisms generally connected with the wave or oscillation amplitude. In all the cases, observed periodicity together with other physical parameters of an observed wave or oscillatory phenomenon (e.g. typical signatures, spatial structures, evolutionary scenarios, etc.) contain information about the physical parameters of the medium. Consequently, waves and oscillations provide us with a natural tool for remote diagnostics of the coronal plasma. This review covers the theory of physical mechanisms for the generation of coronal waves and oscillations, examples of interpretation of recent observational findings in terms of the theory and the use of the results for the remote diagnostics of coronal plasmas.

2 MHD Modes of a Plasma Cylinder

2.1 Dispersion Equation

The standard model for the study of MHD modes of coronal structures is the theory of MHD modes of a straight plasma cylinder. This approach misses several potentially important physical effects, such as the loop curvature, possible smoothness of the loop density and magnetic field profile and twisting of the magnetic field. In addition, in the case of cooler loops, when the scale height is sufficiently small and can be comparable with wave length, the stratification should be taken into account. However, the straight cylinder approach provides us with the possibility to study all its modes analytically which explains its popularity. In the following, we present the formalism developed by [52] and [14].

Consider the oscillating loop as a cylindrical magnetic flux tube of radius a filled with a uniform plasma of density ρ_0 and pressure p_0 penetrated by a magnetic field $B_0 \mathbf{e_z}$; the tube is confined to $r < a$ by an external magnetic field $B_e \mathbf{e_z}$ embedded in a uniform plasma of density ρ_e and pressure p_e. The indices "0" and "e" denote the internal and external media, respectively. Thus, the equilibrium consists of

$$B_0(r) = \begin{cases} B_0, r < a, \\ B_e, r > a, \end{cases} \quad \rho_0(r) = \begin{cases} \rho_0, r < a, \\ \rho_e, r > a, \end{cases} \quad p_0(r) = \begin{cases} p_0, r < a, \\ p_e, r > a. \end{cases} \quad (1)$$

The equilibrium variables are uniform everywhere, except for jumps at the tube boundary $r = a$. Sums of gas and magnetic pressure inside and outside the cylinder are equal to each other to support the total pressure balance condition.

In the internal and external media, the sound speeds are C_{s0} and C_{se}, the Alfvén speeds are C_{A0} and C_{Ae}, and the tube speeds are C_{T0} and C_{Te}, respectively. (The definitions of the speeds are standard, see, e.g., [24]). Relations between those characteristic speeds determine properties of MHD modes guided by the tube.

Linear MHD perturbations of a straight plasma cylinder, evanescent in the external medium, are described by the dispersion relation

$$\rho_e(\omega^2 - k_z^2 C_{Ae}^2)m_0 \frac{I_m'(m_0 a)}{I_m(m_0 a)} + \rho_0(k_z^2 C_{A0}^2 - \omega^2)m_e \frac{K_m'(m_e a)}{K_m(m_e a)} = 0 , \quad (2)$$

where $I_m(x)$ and $K_m(x)$ are modified Bessel functions of order n, and the prime denotes the derivative with respect to argument x. Functions m_0 and m_e which may be considered as radial wave numbers of the perturbations inside and outside the cylinder, respectively, are defined through

$$m_i^2 = \frac{(k_z^2 C_{si}^2 - \omega^2)(k_z^2 C_{Ai}^2 - \omega^2)}{(C_{si}^2 + C_{Ai}^2)(k_z^2 C_{Ti}^2 - \omega^2)} , \quad (3)$$

where $i = 0, e$; For modes that are confined to the tube (evanescent outside, in $r > a$), the condition $m_e > 0$ has to be fulfilled. The integer m is the azimuthal mode number. It determines the azimuthal modal structure: waves with $m = 0$ are *sausage* (or *radial*) modes, waves with $m = 1$ are *kink* modes, waves with higher m are referred to as *flute* or *ballooning* modes. The existence and properties of the modes are determined by the equilibrium physical quantities. In particular, for $m_0^2 < 0$, the internal radial structure of the modes is described by the Bessel functions $J_m(x)$, and the radial dependence of the oscillation is quasi-periodic. The choice of the modified Bessel functions $K_m(x)$ for the external solution is connected with the demand of the exponential evanescence of the solution outside the loop (*trapped modes*).

Figure 1 shows dispersive curves of MHD modes guided by a magnetic cylinder with the characteristic speeds $C_{A0} = 600$ km/s, $C_{Ae} = 3,300$ km/s, $C_{s0} = 340$ km/s and $C_{se} = 200$ km/s. These parameters are typical for a flaring loop. As it has been pointed out in [14], there are fast and slow trapped modes, with longitudinal phase speeds in the intervals $C_{A0} < \omega/k < C_{Ae}$ and $C_{T0} < \omega/k < C_{s0}$, respectively. In the corona, the fast modes are predominantly transverse and the slow modes are predominantly longitudinal. There is an infinite number of modes with a given azimuthal number m, corresponding to different radial number, l. The modes with higher radial numbers l have higher phase speeds along the loop and exist for higher values of the longitudinal wave number k.

In the long wavelength limit $ka \rightarrow 0$, phase speeds of all fast modes except the sausage mode ($m = 0$) approach the kink speed C_k determined as

$$C_k \equiv \left(\frac{2}{1 + \rho_e/\rho_0}\right)^{1/2} C_{A0} . \quad (4)$$

Normally, the kink speed is closer to the Alfvén speed inside the loop. For the physical parameters mentioned above, the kink speed is 900 km/s.

The sausage $m = 0$ mode has a long wavelength cutoff,

$$k_{zc} = \frac{j_0}{a} \left[\frac{(C_{s0}^2 + C_{A0}^2)(C_{Ae}^2 - C_{T0}^2)}{(C_{Ae}^2 - C_{A0}^2)(C_{Ae}^2 - C_{s0}^2)}\right]^{1/2} , \quad (5)$$

where $j_0 \approx 2.40$ is the first zero of the Bessel function $J_0(x)$. Trapped sausage modes can exist only if their longitudinal wave numbers are greater than the

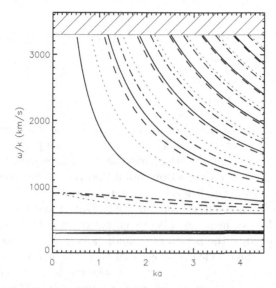

Fig. 1. Dispersion diagram showing the real phase speed solutions of dispersion relation (2) for MHD waves in a magnetic cylinder as a function of the dimensionless parameter $k_z a$. The characteristic speeds in the internal and external media are $C_{A0} = 600$ km/s, $C_{s0} = 300$ km/s, $C_{Ae} = 3300$ km/s and $C_{se} = 150$ km/s. The *solid*, *dotted*, *dashed* and *dot-dashed* curves correspond to solutions with the azimuthal wave number m equal to 0,1,2 and 3 respectively. The torsional Alfvén wave mode solution is shown as a *solid line* at $\omega/k_z = C_{A0}$

cutoff value. For the parameters mentioned above, the sausage mode with the lowest radial number has a cutoff positioned at $ka = 0.51$.

The horizontal axis of Fig. 1 shows the normalised longitudinal wave number, which may be rewritten through the longitudinal wave length $ka = 2\pi a/\lambda$. As fast modes of a coronal loop should have nodes of the transverse velocity perturbations at the loop footpoints, the wave lengths of the modes are quantized, $\lambda = 2L/n$, where n is an integer corresponding to the mode number (or the number of maxima of the transverse velocity perturbations along the loop) and L is the length of the loop. Identification of a particular mode should be based upon the analysis of the mode period and the mode radial and longitudinal numbers. The radial mode number l is perhaps the most difficult to measure, as the loop width is usually either unresolved or resolved very poorly. Fortunately, as modes with higher radial numbers ($l > 1$) can have only short longitudinal wave lengths in comparison with the cylinder radius, in the long wave length limit these modes can be excluded from consideration. The lowest $n = 1$ mode is a fundamental or global mode. Global kink ($n = 1$, $m = 1$) modes have been discovered with TRACE EUV imaging telescope [5, 26]. The period of the global kink mode is

$$P_{\text{GKM}} = \frac{2L}{C_k} , \tag{6}$$

where C_k is given by Eq. (4).

2.2 Global Sausage Modes

The *sausage* mode has been traditionally used for the interpretation of the periodicities in the range 0.5–5 s. The estimation of the periodicity has been based upon the assumption that the period of the sausage mode is determined by the ratio of the loop cross-section radius and the Alfvén speed inside the loop (see, e.g. [2]). However, it has been pointed out in [35, 36] that estimation could be used for higher spatial harmonics only. Unfortunately, the subsequent papers usually did not mention this crucial restriction. Also, this estimation could not be applied to the period of *trapped* global sausage modes.

Dispersion relation (2) clearly shows that there can *not* be trapped sausage modes in the long wave length limit $k \to 0$, where k is the longitudinal wave number. Indeed, in this case, the transverse structure of the mode outside the loop is proportional to $\exp(-m_e r)$, with m_e becoming imaginary (e.g., when $k = 0$, $m_e^2 = -\omega^2/(C_{Ae}^2 + C_{se}^2)$, where C_{Ae} and C_{se} are the Alfvén and sound speeds outside the loop, respectively). In the trapped modes, the external wave number m_e must be *real*, corresponding to the evanescent mode structure outside the loop. Thus, in the long wavelength limit, there are *no* trapped sausage modes of the loop. The very existence of *non-leaky* sausage modes requires finite longitudinal wave numbers. Consequently, the estimation of the global sausage mode period P_{saus} as

$$P_{\text{saus}} = 2\pi a/C_{A0} , \tag{7}$$

where a is the loop cross-section radius and C_{A0} is the Alfvén speed inside the loop (which assumes that the wave is plane, or $k \to 0$) is *incorrect*. This expression, possibly with the change of C_{A0} to a speed in the interval between C_{A0} and C_{Ae} ($C_{A0} < C_{Ae}$), can still be applied to higher spatial sausage harmonics or to leaking modes. However, in this case, Eq. (7) is over-simplified as it does not contain the mode number (actually, it can be correct when the mode number n is exactly equal to $L/\pi a$), but does contain the kink speed which is irrelevant to the sausage mode.

Reminding the reasonings first presented in [36], it was recently pointed out [25] that the period P_{GSM} of the global sausage mode of a coronal loop is determined by the following conditions:

$$P_{\text{GSM}} = 2L/C_p , \tag{8}$$

where C_p is the phase speed of the sausage mode corresponding to the wave number $k = \pi/L$, $C_{A0} < C_p < C_{Ae}$. For $k \to k_c$, C_p tends to C_{Ae} from below (see Fig. 1, and for $k \to \infty$, C_p tends to C_{A0} from above. Also, the length of the loop L should be smaller than π/k_c to satisfy the condition $k > k_{zc}$ given by Eq. (5). For a strong density contrast inside and outside the loop, the period of the sausage mode satisfies the condition

$$P_{\text{GSM}} < \frac{2\pi a}{j_0 C_{\text{A0}}} \approx \frac{2.62a}{C_{\text{A0}}} , \tag{9}$$

as the longest possible period of the global sausage mode is achieved when $k = k_c$. We would like to emphasise that the expression (9) is an *inequality*, and that the actual resonant frequency is determined by Eq. (8), provided (9) is satisfied. Combining (9) and (8), we obtain that the necessary condition for the existence of the global sausage mode is

$$L/2a < \pi C_{\text{Ae}}/2j_0 C_{\text{A0}} \approx 0.65\sqrt{\rho_0/\rho_e} , \tag{10}$$

so the loop should be sufficiently thick and dense.

2.3 Leaky Modes and Radial Modes

The long-wavelength sausage mode, with the wave number shorter than the cut-off value given by (5) does not cease to exist. However, its phase speed becomes complex and the mode leaks out the tube. In this case, parameter m_e given by Eq. (3) becomes imaginary and the appropriate solution in the external medium may be rewritten through the Hankel functions $H_m(x)$ instead of the modified Bessel functions $K_m(x)$. The same change should be made in dispersion relations Eq. (2) [52]. In the long wave length limit $k_z a \ll 1$ and under the assumptions $C_{\text{Ae}} \gg C_{se}$, $C_{\text{A0}} \ll C_{se}$ and $a^2/L^2 \ll \rho_e/\rho_0 \ll 1$, Zaitsev and Stepanov [52] obtained the following estimation for the period of the standing sausage ($m = 1$) or (*radial*) mode:

$$P_{\text{rad}} \approx a/\sqrt{C_{s0}^2 + C_{\text{A0}}^2} . \tag{11}$$

The efficiency of the leakage is apparently connected with the depth of the effective potential well for trapped magnetoacoustic modes, determined by the ratio of the external and internal fast magnetoacoustic speeds. In the low-β coronal plasma that ratio can be rewritten as the ratio of the external and internal plasma densities. The decrement of the sausage mode is

$$\gamma_{\text{leak}} \propto \rho_e/\rho_0 . \tag{12}$$

When the density contrast ratio is small $\rho_e \ll \rho_0$, the leakage can be small as well and, consequently, the oscillations can be of high quality.

2.4 Dispersive Evolution of Fast Wave Trains

In a dispersive medium, impulsively generated (or broadband because of another reason) waves evolve into a quasi-periodic wave train with a pronounced period modulation. In the coronal context, it was pointed out by [14] and [35, 36] that periodicity of fast magnetoacoustic modes in coronal loops is not necessarily connected either with the wave source or with some resonances, but can also be created by the dispersive evolution of an impulsively generated signal. Studying the dispersive evolution, [36] qualitatively predicted that the development of the propagating sausage mode pulse forms a characteristic quasi-periodic wave train with three distinct phases. Such evolution scenario is determined by the presence of minimum in the group speed dependence upon the wave number. An estimation of the generated period, provided in [36], is

$$P_{\mathrm{prop}} = \frac{2\pi a}{j_0 C_{A0}} \sqrt{1 - \frac{\rho_e}{\rho_0}} , \qquad (13)$$

where $j_0 = 2.40$. In loops with the large contrast ratio $\rho_e \ll \rho_0$, Eq. (13) reduces to $P_{\mathrm{prop}} \approx 2.6a/C_{A0}$. We would like to stress that this period should be much shorter than the resonant periods of global modes, otherwise the wave train does not have sufficient distance to get developed. Consequently, this mechanism can operate in sufficiently long and thin loops only.

The analysis presented above is restricted to the case of the slab with sharp boundaries and to sausage modes only. Wave trains with signatures qualitatively similar to the theoretically predicted were found between 303 and 343 MHz data recorded by the digital Icarus spectrometer [36].

The transverse density profile in the loop can affect the fast wave train signature. This was studied [27] by taking the profile as the function

$$\rho_0 = \rho_{\mathrm{max}} \operatorname{sech}^2 \left[\left(\frac{x}{w} \right)^p \right] + \rho_\infty , \qquad (14)$$

where ρ_{max}, ρ_∞ and w are constant. Here, the parameter ρ_{max} is the density at the center of the inhomogeneity, ρ_∞ is the density at $x = \infty$ and w is a parameter governing the inhomogeneity width. The power index p determines the steepness of the profile. The cases when the power index p equals to either unity or infinity correspond to the symmetric Epstein profile or to the step function profile, respectively. In the zero plasma-β case, both profiles give known analytical solutions in the eigenvalue problem. It was established that the qualitative dispersive properties depend weakly upon the specific profile of the density. The group speed has a minimum for all profiles with the power index greater than unity, which are steeper than the symmetric Epstein profile. Thus, the steepness of the profile affects the shape of the wave train and consequently the analysis of wave trains can give us information about this profile.

Numerical simulations of the developed stage of the dispersive evolution of a fast wave train in a smooth straight slab of a low plasma-β plasma [28] showed that development of an impulsively generated pulse leads to formation of a quasi-periodic wave train with the mean wavelength comparable with the slab width. In agreement with the analytical theory, the wave train has a pronounced period modulation which was demonstrated with the wavelet transform technique (Fig. 2). In particular, it is found that the dispersive evolution of fast wave trains leads to the appearance of characteristic "tadpole" wavelet signatures (or, rather a "crazy tadpole" as it comes tail-first).

Rapidly propagating short-period compressible disturbances have recently been discovered with the SECIS instrument during a full solar eclipse [47, 48]. The waves were observed to have a quasi-periodic wave train pattern with a mean period of about 6 s. As the observed speed was estimated at about 2,100 km/s, the propagating disturbances were interpreted as fast magnetoacoustic modes. The comparison of the observed evolution of the wave amplitude along the loop with the theoretical prediction [10] demonstrated an encouraging agreement. The "crazy tadpole" wavelet signatures similar to those shown in Fig. 2 were found in SECIS data [18].

The effect of the dispersive formation of the wave train signature opens up interesting perspectives for MHD coronal seismology. The measurable properties of the wavelet tadpoles are the rates of the frequency and amplitude modulation, determined by the initial spectrum of the wave train, the distance of the region of observation from the wave source and by the loop profile. Multi-point observations can be used to exclude the first two parameters, providing us with the information about the loop profile, its steepness, possible sub-resolution structuring and filling factors.

2.5 Modulation of Gyrosynchrotron Emission by a Loop MHD-oscillations

Another important issue is the modulation of observed emission by the variation of the plasma parameters perturbed by MHD waves. In particular, broadband microwave bursts are generated by the gyrosynchrotron emission mechanism which is very sensitive to the magnetic field in the radio source. Causes of microwave flux pulsations with periods $P \approx 1$–20 s are believed to be some kind of magnetic field variations that modulate the efficiency of gyrosynchrotron radiation or electron acceleration itself. The intensity of optically thin gyrosynchrotron emission at a frequency f is connected with the absolute value of the magnetic field, B and the angle between the magnetic field and the line-of-sight, θ by Dulk & Marsh's approximated formula [13]

$$I_f \approx \frac{BN}{2\pi} \times 3.3 \times 10^{-24} \times 10^{-0.52\delta} (\sin \theta)^{-0.43+0.65\delta} \left(\frac{f}{f_B}\right)^{1.22-0.90\delta}, \quad (15)$$

where N is the concentration of the nonthermal electrons with the energies higher than 10 keV, f_B is the gyrofrequency and δ is the power law spectral index of the electrons. The quasi-periodic variations of the value B and direction of the magnetic field θ can be associated with MHD oscillations of coronal loops (Alfvén and fast modes).

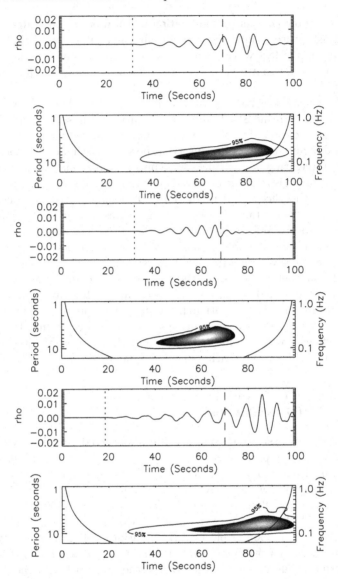

Fig. 2. Numerical simulation of an impulsively generated fast magnetoacoustic wave train propagating along a coronal loop. *Axes labelled 'rho':* Characteristic time signatures of the wave train at the distance $z = 70w$, where w is the loop semi-width, from the source point. The *vertical lines* show the pulse arrival time if the density was uniform; the *dotted line* using the external density, and the *dashed line* the density at the centre of the structure. *Axes labelled 'period':* Wavelet transform analysis of the above signal, demonstrating the characteristic "tadpole" wavelet signature. **Two top panels:** The density contrast ratio is 14.3 and the profile steepness power index equals to 8. **Two middle panels:** The same but for the density contrast ratio equals 5. **Two bottom panels:** The same but for the density profile power index equal to 0.7. (From [28])

3 Standing Longitudinal Waves

Pulsations with longer periods, greater than 20–30 s, can be associated with the slow magnetoacoustic mode which propagates almost along the magnetic field. The perturbations of the field, produced by this mode are usually very weak and the wave can well be described by a one dimensional acoustic approximation. In this approach, the weakly dispersive nature of those modes, connected with the structuring of the plasma, is missing.

Recently, Doppler shift and intensity oscillations were discovered in coronal loops observed in the far UV Fe XIX and Fe XXI emission lines with SOHO/SUMER [19, 44]. This spectral line is associated with temperature of about 6 MK, corresponding to the sound speed of about 370 km/s. The observed periods are in the range 7–31 min, with decay times 5.7–36.8 min, and show an initial large Doppler shift pulse with peak velocities up to 200 km/s. The intensity fluctuation lags the Doppler shifts by 1/4 period.

Initially, these oscillations were interpreted as kink modes. However, after thorough consideration, this interpretation has been excluded since, firstly, it does not explain the observed periodic perturbations of the emission intensity. Indeed, the kink modes are practically incompressible, so the density perturbation, usually responsible for the intensity perturbations, are negligible. But, more importantly, the observed Doppler velocity amplitudes (e.g. 200 km/s), and observed periodicities (e.g. 600 s), together would lead to the kink displacement equal to, e.g. $200 \times 600/2 = 60,000$ km, comparable with the loop length. This definitely cannot be true, and the mode was identified [34] as the acoustic (or longitudinal) one. Indeed, the observed behaviour is consistent with the global standing acoustic mode,

$$V_z(s,t) \propto \cos\left(\frac{\pi C_s}{L}t\right) \cos\left(\frac{\pi}{L}s\right) , \qquad (16)$$

$$\rho(s,t) \propto \sin\left(\frac{\pi C_s}{L}t\right) \sin\left(\frac{\pi}{L}s\right) , \qquad (17)$$

where C_s is a speed of sound, A is wave amplitude, L is loop length, and s is a distance along the loop with the zero at the loop top. According to Eq. (16), the oscillation period is given by the expression

$$P_{\text{long}} = 2L/C_s , \qquad (18)$$

and the practical formula for the determination of the period is

$$P/s \approx 13 \times (L/\text{Mm})/\sqrt{(T/\text{MK})} . \qquad (19)$$

The observed $\pi/4$ shift between velocity and density perturbations (see Fig. 3) is also consistent with the observations. Concerning the decay of the oscillations, it was found that because of the high temperature of the loops, the large thermal conduction, which depends on temperature as $T^{2.5}$, leads to rapid damping of the slow waves on a timescale comparable to observations.

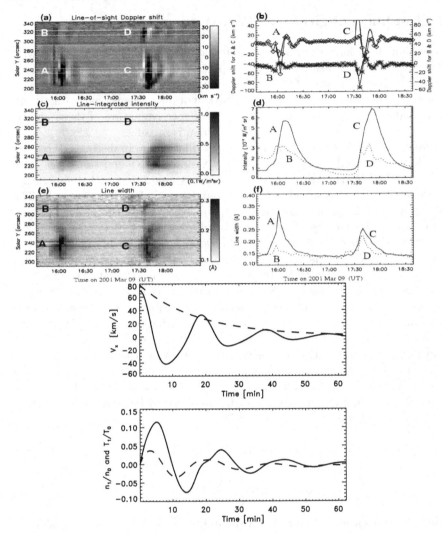

Fig. 3. Top: Doppler oscillation events in the Fe XIX line observed with the SUMER instrument on March 9th, 2001. **(a)** Doppler shift time series. The red shift is represented with the bright colour, and the blue shift with the dark colour. **(b)** Average time profiles of Doppler shifts along cuts AC and BD. The thick solid curves are the best fit functions of the form $V(t) = V_0 + V_D \sin(\omega t + \phi) \exp(-t/t_0)$. **(c)** Line-integrated intensity time series. **(d)** Average time profiles of line-integrated intensities along cuts AC and BD. For a clear comparison, the intensity profile for BD has been stretched by a factor of 10. **(e)** Line width time series. **(f)** Average time profiles of line width along cuts AC and BD. (From [45]) **Bottom**: Temporal evolution of the longitudinal velocity (*top panel, solid line*) and the exponential decay time fit (*dashed line*), produced by a standing acoustic wave in a coronal loop. Perturbations of the density (*lower panel, solid line*) and of the temperature (*dashed line*) at the same location in the loop. (From [34])

This phenomenon can be quite wide-spread in flaring loop dynamics and be observed as X-ray and radio pulsations. The period is determined by the sound speed hence the temperature, the period evolves with loop temperature. Modelling the response of a coronal loop to an impulsive heat deposition (e.g., caused by a flare) [29] demonstrated the effective excitation of the *second* standing harmonics by this mechanism. These oscillations are of high quality and do not experience dissipation. The physical mechanism responsible for the induction of the quasi-periodic pulsations can be understood in terms of auto-oscillations generated by an electric-circuit generator. Indeed, the physical system modelled here contains all the necessary ingredients of a generator: the DC power supply (thermal instability), the nonlinear element (the plasma) and the resonator (the loop). Thus the oscillations may be observed to be dissipationless. However, a proper analytical theory of the excitation of this mode is still to be developed. A currently developed theory of MHD auto-waves and auto-oscillations may be very helpful here. Auto-waves and auto-oscillations are in general a very interesting object for seismological implementations. These dissipative structures are independent of their excitation, but are determined by parameters of the medium only, which makes them an ideal tool for determining those parameters.

The flare-generated second spatial harmonics, producing noticeable perturbations of the loop density, 2–10% of the background, and generating field-aligned flows, may be responsible for observed quasi-periodic pulsations with medium and long periods. For example, [46] observed QPP with the periods of about 50 s at 1.42 and 2 GHz (in association with an M4.4 X-ray flare). Similar periodicities are also detected in the X-ray band, e.g. [21, 41]. The coincidence of QPP periods observed in the X-ray and in radio bands is not a surprise, as the higher frequency radio bursts are found to correlate very well with X-ray bursts (e.g., [6]). Moreover, pulsations with the periods significantly greater than the estimated sausage mode cut-off period have been found in both hard X-ray and microwave bands simultaneously (e.g., 100 s oscillations [15] and 30 s oscillations [43]).

4 Ballooning Modes

The theory of MHD modes of a straight magnetic cylinder misses one very important aspect of coronal loop oscillations - the effects of loop curvature. Indeed, for low-harmonic standing modes the wavelength is greater than the major radius of the loop and consequently the effects of the finite major radius cannot be neglected. In particular, the loop curvature would bring an extra restoring - centrifugal - force, contributing to loop elasticity and hence changing the resonant periods of loop modes. Also, the curvature can make properties of horizontally and vertically polarised waves quite different from each other.

A complete analytical theory of MHD modes of a bent magnetic cylinder has not been created yet. On the other hand, one can use approximate results obtained for laboratory plasma devices, such as tokamaks. In particular, this approach brings important results for the ballooning mode. Consider a loop of a semi-circular shape with the major radius R, neglecting effects of gravitational stratification. Inside the loop the gas pressure is p_0 and the density is ρ_0. Suppose that the equilibrium is perturbed and there is a plasma tongue with the scale $L_1 = L/n$, where $L \approx \pi R$ and n is the number of plasma tongues *along the loop* (or an integer corresponding to the mode number, see Sect. 2.1).

Linear perturbations of this kind are described by the dispersion relation

$$\omega^2 - k_z^2 C_{A0}^2 = -\frac{p_0}{R\rho_0 l} ; \qquad \text{where } l = \begin{cases} b, & b \gg \lambda_\perp , \\ \lambda_\perp, & b \ll \lambda_\perp , \end{cases} \tag{20}$$

where $b = \rho_0 \, (\mathrm{d}/\rho_0 \mathrm{d}r)^{-1}$ is the typical scale of plasma density inhomogeneity across the magnetic field in the loop and $\lambda_\perp$ is the transverse scale of plasma tongue (see, e.g. [23]). Both lengths are restricted by the loop minor radius a.

According to [40], the period of ballooning oscillations of a coronal loop is given by the expression

$$P_{\text{ball}} \approx \frac{L}{C_{A0}} \sqrt{\frac{1}{n^2 - L\beta/(2\pi l)}} . \tag{21}$$

The second term in the denominator, responsible for the ballooning effect, becomes significant for lower-n harmonics when $\beta > 2\pi l/L = l/R \approx a/L$. Otherwise, Eq. (21) reduces to the expression derived in the straight cylinder approximation,

$$P_{\text{ball}} \approx \frac{L}{C_{A0}n} , \tag{22}$$

Expression (21) does not take into account the effect of the external medium, which may modify the resonant period and may cause wave leakage.

For a typical coronal loop, $L = 10^{10}$ cm and $\beta \leq 0.1$, $L\beta/(2\pi l) \leq 1$ justifying the use of Eq. (22).

5 Damping

MHD modes of coronal loops are subject to decay because of several mechanisms. It can be leakage in the corona, discussed in Sect. 2.3, leakage through the footpoint to the photosphere (which is insignificant e.g. for global kink oscillations [31], but may be quite strong for waves with shorter wavelengths) and dissipative processes inside the loop.

According to [40], for radial and ballooning modes, the dominating dissipative mechanisms are electron thermal conductivity with the approximated decrement

$$\gamma_c = \frac{1}{3} \frac{M}{m} \frac{\omega^2}{\nu_{ei}} \beta^2 \cos^2\theta \sin^2\theta \, , \tag{23}$$

and ion viscosity with the decrement

$$\gamma_\nu = \frac{1}{12} \sqrt{\frac{M}{m}} \frac{\omega^2}{\nu_{ei}} \beta \sin^2\theta \, , \tag{24}$$

where m and M are the electron and ion masses, respectively, θ is the angle between the magnetic field and the local wave vector, and ν_{ei} is the effective frequency of electron-ion collisions,

$$\nu_{ei} = \frac{5.5 n_e}{T^{3/2}} \ln\left(10^4 \frac{T^{2/3}}{n_e^{1/3}}\right) \approx 60 \frac{n}{T^{3/2}} \, , \tag{25}$$

where T is the plasma temperature and n_e is electron concentration.

For purely radial oscillations [20], $\theta = 0$ and, consequently thermal conduction is insignificant. For higher longitudinal harmonics or for shorter and thicker loops this approximation cannot be justified. Estimating the ratio of the decrements (23) and (24) we obtain

$$\frac{\gamma_\nu}{\gamma_c} \approx \frac{4 \times 10^{-3}}{\beta \cos^2\theta} \, . \tag{26}$$

For a typical $\beta = 0.1$, we find that the ion viscosity becomes less important than the thermal conduction for the angles $\theta < 80°$.

As the decay rate contains the independent information about the plasma temperature T and the concentration n_e, this observable parameter can be used as a diagnostic tool. Consider QPP in a solar flare observed at 17 GHz and 34 GHz (e.g., with NoRH), assuming that the microwave emission is optically thin and caused by non-thermal electrons (see Sect. 2.5). In this case, the emission flux is $F \propto B^\xi$, where $\xi = 0.9\delta - 1.22$. Assuming that QPP are caused by the modulation of the absolute value of the magnetic field, δB associated with compressible MHD waves (e.g., with a radial mode), we can estimate the modulation depth F produced by this mechanism as

$$\Delta = \frac{F_{\max} - F_{\min}}{F_{\max}} = 2\xi \frac{\delta B}{B} \, . \tag{27}$$

According to Eqs. (23) and (26), for sufficiently short longitudinal wavelengths, $\theta < 80°$ the quality of the oscillations is determined by thermal conduction,

$$Q = \frac{\omega}{\gamma_c} \approx \frac{2m}{M} \frac{P\nu_{ei}}{\beta^2 \sin^2 2\theta} \, . \tag{28}$$

Consequently, the modulation depth, the period and the quality of QPP observed in the microwave band can provide us with a useful tool for MHD coronal seismology.

With the use of expression (22) for the period of ballooning modes, expression (28) for the quality of oscillations, expression (27) for the modulation depth (under the assumption that the emission is produced by the optically thin gyrosynchrotron emission modulated by the perturbations of the magnetic field only) and introducing the parameter $\varepsilon = \Delta/\xi$, we obtain [40] the following relations of physical parameters in the flaring loop of the length L and the QPP's observables ε, P and Q:

$$T = 2.42 \times 10^{-8} \frac{L^2 \varepsilon}{n^2 P^2} \,, n_e = 5.76 \times 10^{-11} \frac{Q L^3 \varepsilon^{7/2} \sin^2 2\theta}{n^3 P^4} \,,$$

$$B = 6.79 \times 10^{-17} \frac{Q^{1/2} L^{5/2} \varepsilon^{7/4} \sin 2\theta}{n^{5/2} P^3} \,. \tag{29}$$

Here the distances are measured in cm, periods in s, temperature in K, concentration in cm^{-3} and the field in G.

Similarly, using expression (11) for the period of leaking sausage (or radial) modes, it was obtained [40] that

$$T = 1.2 \times 10^{-8} \frac{\tilde{r}^2 \varepsilon}{\chi P^2}, n_e = 2 \times 10^{-11} \frac{Q \tilde{r}^3 \varepsilon^{7/2} \sin^2 2\theta}{\chi^{3/2} P^4} \,,$$

$$B = 2.9 \times 10^{-17} \frac{Q^{1/2} \tilde{r}^{5/2} \varepsilon^{7/4} \sin 2\theta}{\chi^{5/2} P^3} \,, \tag{30}$$

where $\chi = 10\varepsilon/3 + 2$ and $\tilde{r} = 2\pi a/j_0 \approx 2.62a$ with a being the loop minor radius. Here it was assumed that the leakage is insignificant and the mode damping is caused by thermal conduction only.

6 Loop Plasma Diagnostics (with Examples)

6.1 Global Sausage Mode (Solar Event of the 12th of January, 2000)

This event gives a very clear example of the global sausage mode oscillations, demonstrating in particular the applicability of the estimation for the global sausage mode period Eq. (8) [25]. Figure 4 shows Fourier spectra measured at different segments of a off-limb flaring loop observed on January 12, 2000 with NoRH in microwaves. There are two clear spectral peaks at the 14–17 s and at 8–10 s. The low period spectral peak is clearly seen near the loop apex, but is depressed near the footpoints, confirming the global mode structure of the oscillation. The shorter period spectral peak (about 9 s) may be associated with sausage modes of higher spatial harmonics. We would like to stress that the period of the second spatial harmonics is not necessarily to be half the period of the global mode, as the fast modes are highly dispersive (see Fig. 1). The length of the flaring loop analysed in [25] was estimated as $L = 25$ Mm and its width at half intensity at 34 GHz as about 6 Mm. These

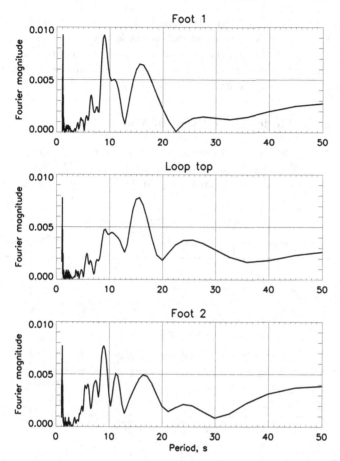

Fig. 4. Fourier power spectra of radio flux variations from different parts of the flaring loop. The upper and lower panels show these flux variations (quasi-periodical pulsations) as received from the 10" × 10" regions at the loop legs. The third panel shows the pulsations from the region of the same size located at the loop apex. Two dominant spectral components with periods $P_1 = 14$–17 s and $P_2 = 8$-11 s are clearly seen for pulsations situated at the different parts of the loop. Note that the longer period component associated with the global sausage mode is more intensive at the apex than in the region close to the footpoints. (From [24])

estimations were confirmed by Yohkoh/SXT images taken on the late phase of the flare. The loop is estimated to be filled by a dense plasma with the electron concentration $n_e \approx 10^{17}$ m^{-3} penetrated by the magnetic field of the strength $B_0 \approx 50-100$ G. According to Eq. (8), the phase speed is $C_p = 3.2 \times 10^3$ km/s. This value is close to and less than the cut-off value $C_p(k_c) = C_{Ae}$. This allows us to estimate the value of the Alfvén speed outside the loop as $C_{Ae} > 3.2 \times 10^3$ km/s. Moreover, from Eq. (9), we get the upper limit on the Alfvén speed inside the loop: $C_{A0} < 5.1 \times 10^2$ km. Thus, this analysis provides us with the estimations of the Alfvén speed values inside and outside the oscillating loop. Assuming that the plasma-β is small and consequently the magnetic field inside and outside the loop has almost equal strength, we can obtain the estimation of the density contrast ratio $\rho_0/\rho_e \approx 40$.

6.2 Ballooning Mode (Solar Event of May 8, 1998)

This M3.1 X-ray class event occurred in the active region NOAA 8210 with coordinates S15 W82 in the time interval 01:49-02:17 UT. Time profiles of impulsive phase of the burst in hard X-ray and microwave emission are shown

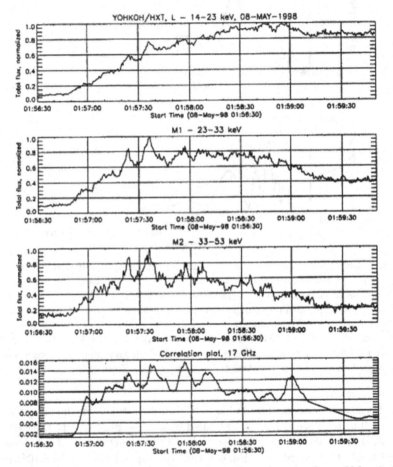

Fig. 5. Left Side: Time profiles for X-ray fluxes from the solar flare of May 8, 1998 in channels L (14–23 keV), M1 (23–33 keV), and M2 (33–53 keV) obtained onboard the Yohkoh satellite, and time profile of 17 GHz burst obtained with the Nobeyama Radioheliograph. **Right Side**: Images of hard X-ray source in channels L and M1. (From [40]) (continued)

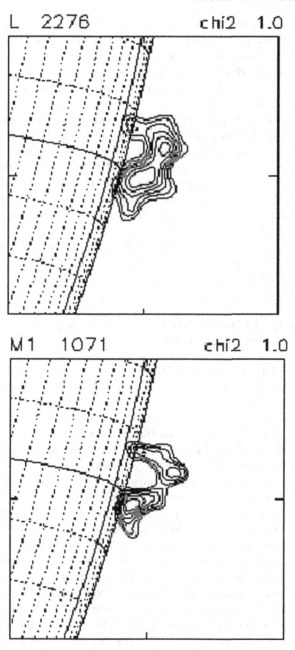

Fig. 5. (continued)

in the left column of Fig. 5. One can see that there is no evident time delay between hard X-ray and 17 GHz pulses. The right column of Fig. 5 shows that in L and M1 channels the source has a "tongue-shape" form, which is typical for ballooning disturbances. It was suggested [40] that the microwave and hard X-ray QPP could be connected with oscillations of plasma tongues - the ballooning mode.

The right panel of Fig. 5 provides some evidence of the presence of four tongues (n = 4) in the loop. The loop length is estimated as $L \approx 8 \times 10^9$ cm. The Fourier analysis of the time profile of the impulsive phase of the flare (Fig. 5, left panel) gives the typical period $P \approx 16$ s. Modulation depth of optically thin gyrosynchrotron emission is $\Delta \approx 0.3$, and the quality $Q \approx 25$. With the use of the thick target X-ray emission model, it is possible to estimate the spectral index of the power-law electrons $\delta = 4.5$, giving $\Delta/\xi \approx 0.11$. The angle θ was estimated as about 66°.

With the use of Eq. (29), Stepanov et al. [40] estimated the temperature $T \approx 10^7$ K, the density $n_e \approx 4.3 \times 10^{10}$ cm^{-3}, and the magnetic field $B \approx$ 230 G in the flaring loop.

6.3 Ballooning and Radial Oscillations of a Loop (Solar Flare of August 28, 1999)

The impulsive phase of this M2.8 X-ray class flare was observed by NoRH in the time interval 00:55-00:58 UT in the active region NOAA 8674 with coordinates S25 W11 [49].

The time profile of the radio flare shown in Fig. 6 reveals QPP. The maximum amplitudes correspond to the three main branches of pulsations with the periods of about 14 s, 7 s, and 2.4 s. Stepanov et al. [40] suggested that the process of the flare energy release could be accompanied by the coalescence of two neighboring loops through the development of ballooning instability in the compact loop. Indeed as is seen in the bottom panel of Fig. 6 the oscillations with the period about 14 s, which can be identified with the ballooning mode, has a time gap (00:55:45 - 00:56:30 UT) that coincides with the onset of propagation of the energetic electrons along the extended loop (see the flare light curve line). It would be natural to attribute this feature to a rise in the gas pressure and to the violation of oscillation conditions in the compact loop, which led to the development of ballooning instability and the injection of both hot plasma and energetic particles into the extended loop. As soon as the compact loop was liberated from the excess pressure, the oscillations of plasma tongues resumed (Fig. 6, bottom). We consider the 7 s oscillations as the second harmonic of the ballooning oscillations. Since the oscillations with the period about 2.4 s emerged only after injection the plasma and energetic particles into the large loop (Fig. 6, bottom), Stepanov et al. [40] concluded that those oscillation were most likely associated with the radial mode. The

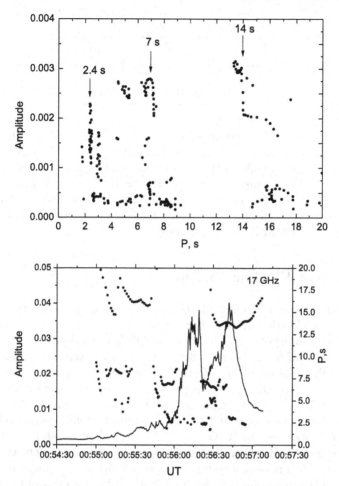

Fig. 6. Top: The dynamic spectrum of oscillations in the 17 GHz flux constructed with the use of wavelet analysis for the solar flare of August 28, 1999. **Bottom**: The time profile of the emission obtained with the Nobeyama Radioheliograph, and the time variations of the oscillation period. (From [40])

process of loop-loop interaction in this flare looks very similar to the Hanaoka event [17].

The period of ballooning oscillations for the fundamental mode (n=1) of compact loop is about $P = L/C_{A0}$ where $L \approx 2 \times 10^9$ cm. For the extended loop with the radius $a \approx 3 \times 10^8$ cm it was obtained that $\tilde{r} = 2.62a \approx 7.8 \times 10^8$ cm. From the observation data we find the quality $Q = 10$ and 15, the modulation depth $\Delta = 0.4$ and 0.1 and the spectral indices $\delta = 5.5$ and 4.0 for the ballooning and radial modes, respectively [49]. With the use of Eqs. (29) and (30), for both the compact and the extended loops Stepanov et al. [40] made the following estimations for the plasma parameters:

Parameter	Extended loop	Compact loop
T, K	2.5×10^7	5.2×10^7
n_e, cm^{-3}	1.0×10^{10}	4.8×10^{10}
B, G	150	280
β	0.04	0.11

In this event and in the event of May 8, 1998 considered above, the plasma density and temperature could be independently estimated from GOES soft X-ray data [40]. In both flares the plasma temperature was estimated as $T \approx 1.5 \times 10^7$ K and the emission measure $n^2V \approx 10^{49}$ cm^{-3}. Taking the loop volume $V = a^2L$ one obtains the plasma density $n_e \approx 4 \times 10^{10}$ cm^{-3} and 5×10^{10} cm^{-3} for the two flares, respectively. This estimation does not contradict to microwave and hard X-ray diagnostics.

7 Equivalent Electric Circuit

There is some observational indication that coronal loops are twisted and hence support electric currents. For example, there are vertical currents of 3×10^{11} A over sunspots [38]. Proposed by Alfvén & Carlquist [1] the electric circuit analogue of a solar flare still remains an attractive model among the numerous other models of solar flares, giving rise to a number of successful coronal studies [22, 39, 53, 55]. In particular, a solar flare with the release rate $dE/dt \approx 10^{20}$ W can be explained by Ohmic dissipation of a 10^{11}–10^{12} A current with the resistance of 10^{-4}–10^{-2} Ohm. To fulfill such a release of energy, the Spitzer conductivity (10^{-11} for typical parameters of a flaring loop) is insufficient. However, Zaitsev & Stepanov [53] pointed out that: (a) a flare is an essential non-steady-state process and (b) the dominant role in the electric current dissipation is played by the neutral component of the plasma. In this case, the Cowling resistance appears to be 7–9 orders of magnitude higher than the Spitzer resistance.

7.1 Advanced Alfvén-Carlquist Flare Model

Consider a coronal magnetic loop with footpoints imbedded into photosphere and formed by the converging flow of photospheric plasma. This structure can be formed when the loop footpoints locate in the nodes of supergranulation cells. The equivalent electric circuit for such a loop can be represented as three domains (Fig. 7). The loop magnetic field and the associated electric current are generated in region 1 located in the photosphere. In this region $\omega_e \gg \nu_{ea}$ and $\omega_i \ll \nu_{ia}$, where ω_e and ω_i are gyrofrequencies of electrons and ions, respectively, and ν_{ea} and ν_{ia} are the frequencies of electron-atom and ion-atom collisions, respectively. Consequently, the electrons are magnetised, while the ions are dragged by the neutral component of the plasma. The

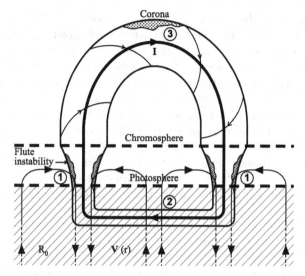

Fig. 7. A cartoon model of current-carrying magnetic loop. Convective plasma flows $V(r) \approx 0.3$–1.0 km/s generate the current in region 1. This current runs through the coronal part of the loop from one foot point to the other and closes in region 2. Flute (*ballooning*) instability in the chromosphere gives high Cowling resistance in the current channel [55]. A "coronal flare" can arise due to filament instability (region 3)

radial electric field E_r is excited due to the charge imbalance, which together with the initial longitudinal magnetic field B_z generates the Hall current j_ϕ and which, in turn, leads to the increase of B_z. The magnetic field grows up to the value when the field enhancement caused by the converging flow is compensated by the magnetic field diffusion because of plasma conductivity. As a result a steady-state flux tube is formed [54]. The electric current runs in the magnetic flux tube from one footpoint to the other and closes in the photosphere (region 2) at the level where the conductivity becomes isotropic. Approximately, the isotropic conductivity corresponds to the level $\tau_{5000} = 1$ (the height of the photospheric region where the optical depth at the wave length 5000 Å is equal to one). Region 3 is the coronal part of the loop. Here plasma $\beta \ll 1$ and the loop magnetic field is force-free, i.e., the electric currents run along the magnetic field lines.

Using the generalized Ohm's law, the Maxwell equations, the equation of motion for the bulk plasma and the mass conservation law one can obtain the rate of Joule dissipation

$$q = \frac{j^2}{\sigma} + \frac{F^2}{c^2 n_e m_i \nu_{ia}} (\mathbf{j} \times \mathbf{B})^2 , \tag{31}$$

where F is the relative density of neutrals and n_e is the concentration of electrons. From Eq. (31) it follows that in force-free magnetic field ($\mathbf{j} \parallel \mathbf{B}$) the current dissipation is due to the Spitzer conductivity. The dissipation is most powerful in the case $\mathbf{j} \perp \mathbf{B}$. The reason for the enhanced current dissipation which can trigger the flare can be the ballooning (flute) instability in the chromospheric part of the loop or in a prominence near the loop top. The instability permits the penetration of partially-ionized plasma tongues into the current channel (Fig. 7), deforming the magnetic field in the loop. As a result, the Ampere force $\mathbf{j} \times \mathbf{B}$ appears and the enhanced current dissipation occurs. A considerable increase of Joule dissipation in partially-ionized gases, first noted by Schlüter and Biermann (1950) [37], is because the energy of the ions dragged through the neutral gas by the Ampere force $\mathbf{j} \times \mathbf{B}$ is much larger than the energy of the relative motion of electrons and ions. Ion-neutral collisions imply great additional dissipation of the energy of electric current. Note, that anomalous resistivity driven by, for example, to give the same effect, the Buneman instability requires stronger filamentation of the current in coronal loop. Therewith cross-sectional area of the filament for $I \approx 3 \times 10^{11}$ A, $n_e \approx 10^{11}$ cm^{-3}, $T \approx 10^7$ K must be $S < I/enV_{Te} \approx 10^{10}$ cm^2. In the case of the ion-sound turbulence the required cross-sectional area is $S < 10^{12}$ cm^2, providing a much softer condition.

7.2 RLC-model: Electric Current Value vs Pulsation Period

The idea that a coronal loop is twisted and then carries an electric current gave rise to several alternative mechanisms for the loop oscillations.

An LCR-circuit model developed by [54] explains the loop oscillations in terms of eigen oscillations of an equivalent electric circuit, where the current I_0 is associated with the loop twist. In this model, small deviations of the electric current $\tilde{I}$ ($\ll I_0$) in the loop are described by the harmonic oscillator equation

$$\frac{1}{c^2}\mathcal{L}\frac{\mathrm{d}\tilde{I}}{\mathrm{d}t^2} + \mathcal{R}_{nl}(I_0)\frac{\mathrm{d}\tilde{I}}{\mathrm{d}t} + \frac{1}{\mathcal{C}(I_0)}\tilde{I} = 0 , \tag{32}$$

where the effective loop capacitance

$$\mathcal{C}(I_0) = \frac{c^4 \rho_0 S^2}{2\pi L I_0^2} \tag{33}$$

is determined mainly by the coronal part of the loop; where c is the speed of light, ρ_0 is the density inside the loop, L and S are the length and cross-sectional area of the coronal part of the loop, respectively; and the loop inductance is

$$\mathcal{L} = 4L\left(\log\frac{8L}{\pi a} - \frac{7}{4}\right) , \tag{34}$$

where a is the minor radius of the loop, $S = \pi a^2$. The period of the electric current oscillations is

$$P_{RLC} = \frac{2\pi}{c}\sqrt{\mathcal{L}\mathcal{C}} . \tag{35}$$

For typical parameters of a flaring loop, $n_e = 10^{10}$ cm^{-3}, $L = 5 \times 10^9$ cm, a practical formula for the estimation of the period is

$$P_{RLC}/\text{s} \approx (I_0/10^{12} \text{ A})^{-1} . \tag{36}$$

Oscillations of the electric current in the loop can modulate thermal and nonthermal emission from the loop.

According to Eq. (34) the loop inductance can be quite large. For $\mathcal{L} \approx 10$ H, the total energy in the loop $\mathcal{L}I_0^2/2c^2 \approx 10^{30}$–$10^{32}$ erg. The analysis of flare events observed at 37 and 22 GHz at Metsähovi showed that only about 5–10% of the electric current energy stored in the loop was released in the flares [54].

7.3 Observational Evidence for Energy Accumulation and Dissipation in Coronal Magnetic Loop

The analysis of solar flares observed in 1990–1993 with the Metsähovi radio telescope at 37 GHz with the use of the Wigner–Ville transform [9] has been performed in [50] (see Fig. 8 for a typical example). This method makes it possible to obtain the dynamical spectrum of low-frequency pulsations, i.e. to determine the time evolution of the frequency. This information allows to identify the types of oscillations observed and to determine their Q-factors and other temporal characteristics. This is very important for the diagnostics of the parameters of the emission source.

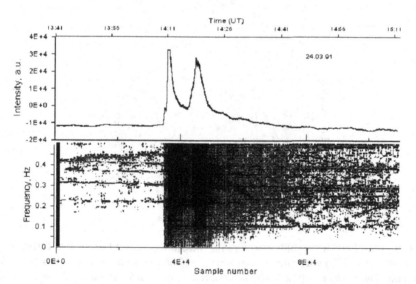

Fig. 8. A solar flare burst on 24th of March, 1991 at 37 GHz (Metsähovi) and dynamic spectrum of low-frequency pulsations obtained using the Wigner–Ville transform [50]

The negative frequency drift clearly seen in Fig. 8 is consistent with the interpretation of the oscillations in terms of the RLC-circuit model. According to Eq. (35), the increase of the period can be produced by the decrease of the current. In particular, the observed frequency shift could be associated with the decrease of the current from 9×10^{11} A to 10^{11} A. This would provide a useful diagnostics for the dissipation of the electric current energy during the flare ($\approx 10^{28}$ erg/s) and of the accumulation of the energy before and after flare.

7.4 Diagnostic of Coronal Loop Plasma on AD Leo (Flare Event of May 19, 1997)

Phenomenologically there is much in common between coronal radio QPP observed on the Sun and on late-type stars [7, 16]. This can be due to the

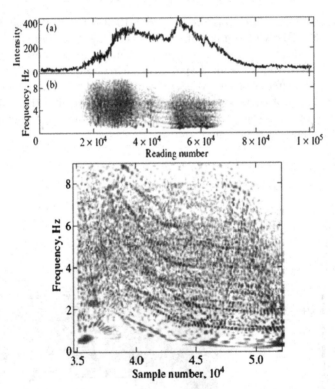

Fig. 9. Top: (a) Time profile of the radio emission from AD Leo at 4.85 GHz (May 19, 1997,18.945 UT) with time resolution 1 ms. Total radio flare duration is 50 s. (b) Dynamic low-frequency (0.2–9 Hz) modulation spectrum for the microwave emission from AD Leo obtained by using Wigner-Ville transform. **Bottom:** A portion of the dynamic low-frequency modulation spectrum obtained for times at the descending branch of the time profile for the first radio pulse. (From [51])

similarities of the physical processes in solar and stellar flares. Therefore the diagnostic methods for solar flare pulsations can be applied for stellar flare QPP as well. Very good examples of stellar coronal QPP are supplied by the radio observations of the active red dwarf AD Leo (Gliese 388, dM3.5e). Figure 9 shows the time profile of the radio emission from this star detected on May 19, 1997 (18.945 UT) at 4.85 GHz with the 100-m Effelsberg radio telescope. The emission flux was about 400 mJy at the burst peak. The time profile shows fluctuations of the radio flux with a modulation depth from several to 10%.

Dynamic spectra of low-frequency pulsations (Fig. 9) obtained with the Wigner-Ville transform reveals nearly equidistant, frequency-drifting narrow-band signals with frequency splitting, the mean frequencies are in the range of 0.5–2.0 Hz [51]. From the analysis it was concluded that both the radial mode ($P \approx 0.5$ s) and RLC-oscillations ($P =0.5$–2.0 s) can explain the observed periodicity. Applying Eqs. (29) and (30) to the diagnostics of the flaring loop, the Alfvén speed was estimated as 3.5×10^8 cm/s, the loop radius as 1.8×10^8 cm, the loop length as 4×10^{10} cm, the current strength as 4.5×10^{12} A, the loop inductance 50 H, the energy stored in the loop 10^{33} erg, and the rate of the energy release 10^{32} erg/s which is 2–3 orders larger than in typical solar flares. This is possible as the AD Leo is believed to have stronger magnetic fields and more active surface convection compared to the Sun.

8 Conclusions

The theory and the observational examples discussed above demonstrate the applicability of the method of solar and stellar coronal seismology in the radio band. MHD coronal seismology is a rapidly developing new branch of coronal physics, so far based upon the use of EUV and visible light observations. The radio band is of particular interest for the seismological efforts, as radio observations usually have sufficiently high temporal resolution comparable with the transverse transit time across coronal loops. This allows for the probing of coronal structures unresolved even with the best present EUV coronal imagers, e.g. TRACE, bringing us information about the transverse structure of coronal loops which are the main building blocks of solar active regions. Also, the mechanisms for the generation of coronal QPP are directly based on the emission modulation by the absolute value and direction of the coronal magnetic field and by the coronal electric currents - the physical parameters crucial for our understanding of coronal physics but not open to the direct observations - which makes this approach especially important.

Solar coronal diagnostics with radio QPP is essentially based upon the use of the observational tools allowing for the spatial resolution, e.g. NoRH. The spatial information is crucial for the determination of the mode type and number and hence the correct mode identification.

The first attempts of coronal diagnostics with radio QPP discussed in the review are very encouraging. However, there is room for further development of this method in its both observational and theoretical aspects. For example, the longitudinal mode, routinely observed in the EUV band, has not been identified in the radio band yet. Kink oscillations, also often detected in EUV, have been applied to the interpretation of radio QPP, but the results are not completely conclusive. The RLC-circuit model of QPP should be connected with the MHD model. The MHD model itself needs to be generalised to self-consistent accounting for the effect of loop curvature. The mechanisms for the modulation of the radio emission by variations of the magnetic field and electric currents in a flaring loop need to be understood, explaining, in particular, the efficiency of the modulation. The links between radio and soft and hard X-ray QPP have to be studied.

All above makes the study of coronal radio QPP an interesting and important research area. The potential of coronal diagnostics with radio QPP should be exploited in full.

References

1. Alfvén, H., Carlqvist, P.: Solar Phys. **1**, 220 (1967)
2. Aschwanden, M. J.: Solar Phys. **111**, 113 (1987)
3. Aschwanden, M. J.: in: *Turbulence, Waves and Instabilities in the Solar Plasma*, ed by R. Erdélyi et al., (Kluwer, 2003) p. 215
4. Aschwanden, M. J.: *Physics of the Solar Corona*, (Springer Praxis Books, Netherlands 2004)
5. Aschwanden, M. J., Fletcher, L., Schrijver, C. J. and Alexander, D.: Astrophys. J. **520**, 880 (1999)
6. Benz, A. O., Kane, S. R.: Solar Phys. **104**, 179 (1986)
7. Bastian, T., Bookbinder J., Dulk G. A., Davis M: Astrophys. J. **353**, 265 (1990)
8. Berghmans, D. and Clette, F.: Solar Phys. **186**, 207 (1999)
9. Cohen, L.: Proc. IEEE **77**, 941 (1989)
10. Cooper, F. C., Nakariakov, V. M., Williams, D. R.: Astron. Astrophys. **409**, 325 (2003)
11. DeForest, C. E. and Gurman, J. B.: Astrophys. J. **501**, L217 (1998)
12. De Moortel, I., Ireland, J. and Walsh, R. W.: Astron. Astrophys. **355**, L23 (2000)
13. Dulk, G. A., Marsh, K. A.: Astrophys. J. **259**, 350 (1982)
14. Edwin, P. M. and Roberts, B.: Solar Phys. **88**, 179 (1983)
15. Fu, Q.-J., Liu, Y.-Y., Li, C.-S.: Chin. Astron. Astrophys. **20**, 487 (1996)
16. Güdel, M., Benz, A. O., Bastian, T. S. et al.: Astron. Astrophys. **220**, L5 (1989)
17. Hanaoka, Y.: Proc. in Nobeyama Symposium NRO No 479, (eds. Bastian T., Gopalswamy N., Shibasaki K.,), 229 (1999)
18. Katsiyannis, A. C., Williams, D. R., McAteer, R. T. J., et al.: Astron. Astrophys. **406**, 709 (2003)
19. Kliem, B., Dammasch, I. E., Curdt, W. and Wilhelm, K.: Astrophys. J. **568**, L61 (2002)

20. Kopylova, Y. G., Stepanov, A. V., Tsap, Y. T.: Astron. Lett. **28**, 783 (2002)
21. McKenzie, D. E. and Mullan, D. J.: Solar Phys. **176**, 127 (1997)
22. Melrose, D. B., McClymont, A. N.: Solar Phys. **113**, 241 (1987)
23. Mikhailovskii, A. B.: Plasma instability theory. Volume 1 - Instabilities in a homogeneous plasma, Moscow, Atomizdat (1975)
24. Nakariakov, V. M.: Coronal oscillations. In: *Dynamic Sun*, ed by B. Dwivedi (Cambridge University Press, Cambridge, 2003) pp. 314–334
25. Nakariakov, V. M., Melnikov, V. F. and Reznikova, V. E.: Astron. Astrophys. **412**, L7 (2003)
26. Nakariakov, V. M., Ofman, L., DeLuca, E. E., Roberts, B. and Davila, J. M.: Sci. **285**, 862 (1999)
27. Nakariakov, V. M. and Roberts, B.: Solar Phys. **159**, 399 (1995)
28. Nakariakov, V. M., Arber, T. D., Ault, C. E., Katsiyannis, A. C., Williams, D. R. and Keenan, P.: MNRAS **349**, 705 (2004)
29. Nakariakov, V. M., Tsiklauri, D., Kelly, A., Arber, T. D. and Aschwanden, M. J.: Astron. Astrophys. **414**, L25 (2004)
30. Nakariakov, V. M., Verwichte, E., Berghmans, D. and Robbrecht, E.: Astron. Astrophys. **362**, 1151 (2000)
31. Ofman, L.: Astrophys. J. **568**, L135 (2002)
32. Ofman, L., Romoli, M., Poletto, G., Noci, G. and Kohl, J. L.: Astrophys. J. **491**, L111 (1997)
33. Ofman, L., Nakariakov, V. M. and DeForest, C. E.: Astrophys. J. **514**, 441 (1999)
34. Ofman, L. and Wang, T.: Astrophys. J. **580**, L85 (2002)
35. Roberts, B., Edwin, P. M., Benz, A. O.: Nature **305**, 688 (1983)
36. Roberts, B., Edwin, P. M., Benz, A.O.: Astrophys. J. **279**, 857 (1984)
37. Schlüter, A., Biermann, L.: Z. Naturforsch. **5a**, 237 (1950)
38. Severny, A. B.: Space Sci. Rev. **3**, 451 (1965)
39. Spicer, D. S.: Solar Phys. **70**, 149 (1981)
40. Stepanov, A. F., Kopylova, Y. G., Tsap, Y. T., Shibasaki, K., Melnikov, V. F., Goldvarg, T. B.: Astron. Lett. **30**, 480 (2004)
41. Terekhov, O. V., Shevchenko, A. V., Kuz'min, A. G., Sazonov, S. Y., Sunyaev, R. A., Lund, N.: Astron. Lett. **28**, 397 (2002)
42. Thompson, B. J., Plunkett, S. P., Gurman, J. B. et al.: Geophys. Res. Lett. **25**, 2465 (1998)
43. Tian, D.-W., Gao, Z.-M., Fu, Q.-J.: Chin. Astron. Astrophys. **23**, 208 (1999)
44. Wang, T. J., Solanki, S. K., Curdt, W., Innes, D. E. and Dammasch, I. E.: Astrophys. J. **574**, L101 (2002)
45. Wang, T. J., Solanki, S. K., Curdt, W., Innes, D. E., Dammasch, I. E. and Kliem, B.: Astron. Astrophys. **406**, 1105 (2003)
46. Wang, M., Xie, R. X.: Chin. Astron. Astrophys. **24**, 95 (2000)
47. Williams, D. R., Phillips, K. J. H., Rudawy, P. et al.: Mon. Not. R. Astron. Soc. **326**, 428 (2001)
48. Williams, D. R., Mathioudakis, M., Gallagher, P. T. et al.: Mon. Not. R. Astron. Soc. **336**, 747 (2002)
49. Yokoyama, T., Nakajima, H., Shibasaki, K., Melnikov, V. M., Stepanov, A. V., Astrophys. J. **576**, L87 (2002)
50. Zaitsev, V. V., Kislyakov, A. G., Urpo, S., Stepanov, A. V., Shkelev, E. I.: Astron. Rep. **47**, 873 (2003)

51. Zaitsev, V. V., Kislyakov, A. G., Stepanov, A. V., Kliem, B., Furst E.: Astron. Lett. **30**, 319 (2004)
52. Zaitsev, V. V., Stepanov, A. V.: Issled. Geomagn. Aeron. Fiz. Solntsa **37**, 3 (1975)
53. Zaitsev, V. V., Stepanov, A. V.: Solar Phys. **139**, 343 (1992)
54. Zaitsev, V. V., Stepanov, A. V., Urpo, S., Pohjolainen, S.: Astron. Astrophys. **337**, 887 (1998)
55. Zaitsev, V. V., Urpo, S., Stepanov, A. V.: Astron. Astrophys. **357**, 1105 (2000)

Pulsating Solar Radio Emission*

Alexander Nindos[1] and Henry Aurass[2]

[1] Section of Astrogeophysics, Physics Department, University of Ioannina,
   Ioannina GR-45110, Greece
   anindos@cc.uoi.gr
[2] Astrophysical Institute Potsdam, D-14482 Potsdam, Germany
   haurass@aip.de

Abstract. A status report of current research on pulsating radio emission is given, based on working group discussions at the CESRA 2004 workshop. Quasi-periodic pulsations have been observed at all wavelength ranges of the radio band. Usually, they are associated with flare events; however since the late 90s, pulsations of the slowly-varying component of the Sun's radio emission have also been observed. Radio pulsations show a large variety in their periods, bandwidths, amplitudes, temporal and spatial signatures. Most of them have been attributed to MHD oscillations in coronal loops, while alternative interpretations consider intrinsic oscillations of a nonlinear regime of kinetic plasma instabilities or modulation of the electron acceleration. Combined radio spectroscopic observations with radio imaging and X-ray/EUV data have revived interest in the subject. We summarize recent progress in using radio pulsations as a powerful tool for coronal plasma and magnetic field diagnostics. Also the latest developments on the study of the physical processes leading to radio emission modulation are summarized.

1 Introduction

The interaction between the plasma and magnetic field produces several dynamic phenomena in the solar corona. Among them, significant attention has been paid to "pulsations" of the observed radiation. In the magnetized solar plasma, the eigen-frequencies of any oscillating system are determined by the magnetic field strength, and also the density and configuration of the magnetized plasma. Consequently, the determination of the oscillation properties helps us to derive information about the plasma and magnetic field parameters that is otherwise unavailable.

Over the years, several reports on temporal oscillations in the corona have been published. They cover almost all wavelengths with the majority in radio wavelengths (e.g. see the reviews by [4, 5]). The oscillations show a large variety in their periods, wavelengths, bandwidths, amplitudes, temporal and

---

* Working Group Report.

A. Nindos and H. Aurass: *Pulsating Solar Radio Emission*, Lect. Notes Phys. **725**, 251–277
(2007)
DOI 10.1007/978-3-540-71570-2_12

spatial signatures. This reflects the fact that pulsating emission can be generated by different physical mechanisms on different time and spatial scales. Most of them have been attributed to MHD oscillations in coronal loops, while pulsations with irregular periodicities have been interpreted in terms of limit cycles of nonlinear wave-particle interactions in coronal loops. Another approach presumes a periodically repeated injection of nonthermal electrons into a coronal loop thus shifting the problem to a periodically acting accelerator.

Until the late 90s most reports on coronal pulsations came from time-series analysis of total flux temporal oscillations. With the development of suitable space-borne and ground-based instruments, the detection of spatially-resolved pulsations has become possible in EUV and microwave wavelengths. This led to the introduction of the term *coronal seismology*, even though the concept is not new: one compares the observed properties of the oscillations with theoretical modeling of the wave phenomena and obtains estimates of the coronal dissipative coefficients (e.g. [46]) and magnetic field (e.g. [47]). Usually, pulsating radio emission is associated with solar flares; however, since the late 90s, quasi-periodic oscillations of the slowly-varying component of the Sun's radio emission (i.e. emission from non-flaring active regions) have also been observed.

The radio domain is well known for its advantages, such as the possibility to observe the corona from the ground with unparalleled temporal resolution, the capability to probe the solar atmosphere from the chromosphere to the middle corona, the relative simplicity of (several) physical processes involved in the radio emission, the density resolution due to plasma emission, and the sensitivity to the magnetic field in the source volume. Of course, equally well known are its inherent limitations, for example the inferior spatial resolution and the influence of refraction effects and scattering. The above statements together with the prevalence of observations of radio pulsations in the literature justify the existence of the *Working Group on Radio Pulsations* at the CESRA Workshop on *The High Energy Solar Corona: Waves, Eruptions, Particles*.

In this report we attempt to capture the content of discussions held in the working group's sessions. In order to put the presentation of the issues on a concrete footing, in Sect. 2 we present a brief overview of radio pulsations. A detailed review of the subject appears in the article by Nakariakov and Stepanov in this volume. In Sects. 3, 4, and 5 we present the issues discussed in the working group's sessions after we review –definitely not thoroughly– the existing literature. A summary and suggestions for future work are given in Sect. 6. Our report follows the traditional division into spectral ranges: microwave, decimetric and metric. This is certainly arbitrary in terms of the underlying physics responsible for the radio events but it is justified mainly by the fact that a self-consistent picture covering all aspects of radio pulsations in the solar corona is still missing. Although our working group did not solve any problems, we believe that we did make some progress in putting together some of the observational data and pointing out outstanding questions.

2 A Brief Overview of Radio Pulsations

2.1 Pulsations Associated with Transient Activity

Quasi-periodic pulsations of coronal radio emission have been observed extensively for many years. In solar radio astronomy, the term "pulsation" sometimes is used liberally to describe single-frequency total flux time profiles showing repeated ups and downs quasi-periodically. For early examples, the reader is referred to [20], who reported continuum fluctuations of noise storms and type IV bursts with minute time-scale, and [19] who first described decimetric pulsations. This definition does not tell anything about the spectral bandwidth of the fluctuations and may contribute to significant ambiguity in the interpretation of observations. For example, groups of type III bursts, type II burst herring-bone fine structures, and so called broadband pulsations can all appear as quasi-periodic oscillations on single frequency records despite the fact that they reflect completely different physical processes.

Consequently, additional spectral information is needed when studying periodic fluctuations of the total radio flux at discrete frequencies. The first spectral data showing metric broad band pulsations were presented by [74]. For decades, radio spectra were recorded on film. Now the digital recording and processing of broad band receiver data gives much easier access to the data, especially to fast intensity fluctuations with time and frequency.

Radio pulsations have been observed at all wavelength ranges of the radio band: metric, decimetric and microwaves. They demonstrate a rich diversity of observational features. Periods from 10^{-2} to 10^3 s have been reported. The pulse amplitude and pulse periodicity range from almost constant to rapidly varying. Broad-band (e.g. $\Delta f/f \approx 1$) and narrow-band (e.g. $\Delta f/f \leq 0.1$) pulsations have been detected.

Typical examples of radio pulsations are given in the dynamic spectra of Fig. 1 (narrow-band pulsations in Fig. 1b and broad-band pulsations in Fig. 1a). Broad-band pulsating structures are a well-known fine structure phenomenon appearing in dynamic spectra of solar type IV continuum emission. Their characteristic period is of the order of 1 s (e.g. [1, 43, 62]). Pulsating sources were first imaged with the Japanese 160 MHz interferometer [29], and later with the French Nançay Radio Heliograph (NRH), see [69]. Also a variety of other patterns can be superposed on metric and/or decimetric continua: intermediate drift or fiber bursts, with bandwidths below 10 MHz, durations of less than 1 s and drift rates between the fast (type III) and slow (type II) drift bursts. Furthermore, zebra patterns of almost parallel, drifting emission bands and spike bursts can occur as superposed continuum fine structures.

Fiber bursts, spikes, and zebra patterns do not belong to periodic pulsations in the strict sense of the term. But sometimes all these phenomena cannot be distinguished easily for two reasons: first, they cannot be discriminated with single frequency records alone, and second, all these fine structures

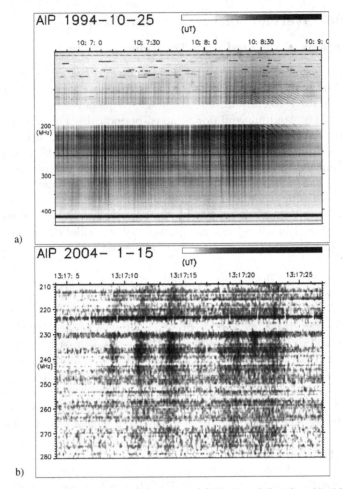

Fig. 1. Examples of meter-wave pulsations. (**a**) Event of October 25, 1994, complex type II/IV burst [8]. Broad-band pulsations with more than 100 single pulses, followed–at the low frequency edge–by more than 20 parallel *zebra stripes*. The *zebra stripes* are sometimes already visible between the pulsation pulses. The low frequency limit in Fig. 1a is 110 MHz. (**b**) Event of January 15, 2004. Faint isolated narrow-band pulsation event of 5 pulses only, but with the same period as in (**a**). Their comparison with Trieste single frequency records confirmed their solar origin (courtesy J. Magdalenić)

can occur, at least in the metric range, in continuum patches together (with the term "continuum patches" we mean structures with duration too long for a type III burst and too short for a type IV burst), sometimes seemingly in a systematic sequence in time. Therefore a common source mechanism has been invoked (e.g. [36]). Because of this situation, we did not exclude the above-mentioned fine structure phenomena from the present article about pulsating radio phenomena.

The pulsating radio emission can come from either a coherent or incoherent emission mechanism. When the pulsations show narrow spectral bandwidth, a coherent mechanism should be involved. The physical mechanisms producing quasi-periodic pulsations have been grouped into three categories e.g. by [36] and [4].

1) MHD oscillations of a magnetic loop which modulate the radio emission. There are three branches of solutions of the dispersion relation for propagating and standing MHD waves: the slow-mode branch (with acoustic phase speed) and the fast-mode and Alfvén branch (both with Alfvénic phase speed). Each branch has a symmetric and asymmetric solution, named the sausage and kink modes [57]. The fast kink mode produces a lateral translation of the loop but does not modify magnetic field strength or particle density. Consequently, it is difficult to modulate radio emission with the kink mode. The longitudinal (slow magnetoacoustic) mode yields periods much longer than the ones usually observed in radio. Numerous nonimaging radio observations with periods of about 0.5 s up to some minutes have been interpreted in terms of the fast sausage mode. Ideally, it is a mode in which a slender flux tube oscillates via standing symmetric fast mode waves [56]. Fast mode waves can propagate only if the Alfvén speed in the loop is slower than in the surroundings; i.e. the loop must be overdense. If we assume a cylindrical symmetry, the period is determined by the length of the loop (e.g. [49]).

The generation of MHD oscillations is usually associated with either the initial energy release in the dense kernel of the flare [76], or with plasma flows induced by chromospheric evaporation [77]. Alternative mechanisms are the particle acceleration in current sheets [4] and in colliding current-carrying loops [60], and in a single current-carrying loop considered as an LCR-circuit [78].

2) Intrinsic oscillations of the flux created by an oscillatory nonlinear regime of the kinetic plasma instabilities that emit the radio waves. Oscillatory wave-wave interactions (e.g. [75]) and wave-particle interactions (e.g. [64]) have been considered. The common ingredient of these mechanisms is that the pulsating regime corresponds to the limit cycle in phase space of a nonlinear dissipative system, and thus may show smaller (for linear disturbances) or larger (for highly nonlinear conditions) deviations from strict periodicity.

3) Modulation of acceleration. Here, one idea is that the pulsating radio emission comes from a resonant region which is driven by external dynamics. An example is the model by [34] which has been used for the interpretation of some decimetric pulsations: the model invokes quasi-periodic particle acceleration episodes that result from a dynamical regime of magnetic reconnection in a large-scale current sheet. This process leads to the formation of a growing plasmoid, which becomes strongly accelerated along the sheet, and to a pulsating particle source at the magnetic X line adjacent to the plasmoid.

Alternatively, referring to the analysis of X-ray and microwave observations by [26], the interaction between a small-scale emerging loop hosting the particle accelerator, and a large-scale loop being a trap for injected particles and containing the pulsating radio source, has been considered. Such an approach sometimes helps us understand the excitation of meter-wave pulsations [79].

2.2 Quasi-periodic Oscillations of the Microwave Slowly-varying Component

The fact that the atmosphere above sunspots shows modes of oscillatory behavior, visible as intensity and velocity variations, is well known for several decades (e.g. see the reviews by [39], and [63]). In the umbral photosphere, oscillations with periods in the five–minute range as well as in the three–minute range occur. At chromospheric levels the intensity and velocity oscillations with periods of 150–200 s show larger amplitudes and are observed in the inner part of the umbra. Reports of photospheric magnetic field oscillations also exist (e.g. [52, 59]). However, it seems that, at least in some of the reported observations of photospheric magnetic field oscillations, instrumental effects have an important contribution to the observed oscillatory pattern (e.g. [11]). Observations with instruments on board SOHO (Solar and Heliospheric Observatory) have confirmed the presence of mostly 3-minute oscillations in the coronal-chromosphere transition region (TR) above sunspots (e.g. [17, 22]). Usually, they affect the entire umbral TR and part of the penumbral TR. Most of them are compatible with the hypothesis that the oscillations are caused by linear, upward–propagating progressive acoustic waves (however, [18] reported nonlinear oscillations in the TR of one sunspot).

Since the late 90s, in addition to the microwave pulsations associated with flare activity, quasi-periodic oscillations of the slowly varying component of the microwave emission of active regions have been detected ([23]; see also [24]). These pulsations are observed primarily above sunspots and, in principle, are independent of any flare activity. The sunspots involved emit strong stable gyroresonance radiation. [23] used Nobeyama Radioheliograph (NoRH) data at 17 GHz and measured the circular polarization (V) flux of several such sources as a function of time; they detected nearly harmonic oscillations with periods mostly between 120–220 s.

In order to discuss the origin of these oscillations we briefly remind the reader of the properties of the gyroresonance emission mechanism (for more details see, e.g., the review by [71]). Gyroresonance is a resonant mechanism: opacity is significant only in thin layers where the observing frequency is a low–integer multiple of the local gyrofrequency. The gyroresonance emission in the extraordinary (x) mode comes primarily from the third or lower harmonics of the gyrofrequency, whereas ordinary (o) mode emission has less opacity and comes from the second harmonic. The structure of the source depends upon which of the low–order (second to third; the fourth may play some role in big sunspots very close to the limb) harmonics of the gyrofrequency are located in regions of high temperature. This, in turn, depends on the frequency and the magnetic field. The brightness temperature of a harmonic that satisfies the above condition is determined by the electron temperature at the height where it is located and by its opacity. The opacity has much stronger dependence on the angle θ between the magnetic field and the line of sight than on any other physical parameter such as temperature and density. Gelfreikh et al. [23] interpreted the periodic fluctuations of the V flux they detected in terms of MHD oscillations which result in variations of the size of the emitting gyroresonance layer and its temperature.

Furthermore, [61] detected 3–min oscillations in the 17 GHz emission of a sunspot for which 3–min velocity and intensity oscillations due to upward–travelling acoustic waves had been detected in TR lines observed by SUMER. Shibasaki [61] applied the values of density and temperature fluctuations deduced from SUMER to the sunspot's gyroresonance emision and found good agreement with the detected radio oscillation. He attributed the 3–min oscillation to the resonant excitation of the cut–off frequency mode of the temperature plateau around the temperature minimum. Using the Very Large Array (VLA) at 8.5 and 5 GHz, [50] were able to detect spatially resolved oscillations in both the total intensity and circular polarization emission of a sunspot-associated gyroresonance source (see Sect. 3.2 for more details).

3 Microwave pulsations

3.1 Flare-related Microwave Pulsations

Microwave bursts often display quasi-periodic pulsations with periods from about 40 m sec in narrow-band bursts (e.g. [21]) up to 20 sec in broad-band bursts (e.g. [3, 25, 45, 65, 55]). While the narrow-band bursts require a coherent mechanism to be involved, the broad-band bursts are generated by the gyrosynchrotron emission mechanism which is very sensitive to the magnetic field in the radio source. Broad-band microwave flux pulsations with periods $P \approx 1$–20 s are believed to represent some kind of magnetic oscillations that modulate the efficiency of gyrosynchrotron radiation or electron acceleration itself (see Sect. 2.1).

Fleishman (see also [21]) presented two events showing periodic narrow-band millisecond pulsations of the microwave emission both in total intensity and circular polarization. He found unusually large delays between the right and left-hand circularly polarized radiation, and showed that the radio emission was generated as unpolarized by a plasma mechanism at the second harmonic of the upper-hybrid frequency. His analysis demonstrated that the observed oscillations of the degree of circular polarization could be understood in terms of a group delay between x-mode and o-mode along the line of sight. The predicted theoretical dependence of the group delay with frequency ($\propto f^{-3}$) agreed very well with the observed frequency dependence of the delay between the right-handed and left-handed polarized components of oscillations. Also the authors were able to deduce a number of source parameters in addition to the density and magnetic field strength of the background plasma.

Fleishman also presented a theoretical study which shows that the mechanism of transition radiation (i.e. radiation generated by charged particles as they move through a boundary between two media having different dielectric permeabilities, or when they move through a spatially or temporally inhomogeneous medium) in turbulent dense media [54] yields common microwave and decimetric sources out of the same loop structure and from the same energetic particle ensemble. This is interesting because it provides a link to emission features frequently occurring at both spectral ranges.

Sheiner reported microwave pulsations before strong flares. She derived periods of 5–6 s for pulsations at $\lambda = 3$ cm and periods of 7–9 s at $\lambda = 10$ cm. Careful follow-up is needed in order to check whether instrumental effects do not contribute to the reported oscillatory patterns.

Reznikova and Melnikov reported spatially resolved oscillations in a flaring loop observed at 17 and 34 GHz with NoRH (see also [7, 49]). The total flux integrated over the entire source did not reveal oscillations; the oscillations were apparent at selected $10'' \times 10''$ regions at the footpoints and loop top of the flaring loop (see Fig. 2). The oscillations were interpreted in terms of sausage magnetoacoustic modes. First, they showed that the oscillation period of the global sausage mode is determined by the loop length and the density contrast between the loop and its environment. They noted that the previously used expression for the sausage-mode oscillating period (depending primarily on the ratio of the loop cross-section radius and the Alfvén speed inside the loop) is not correct because it does not take into account the highly dispersive nature of the phase speed and the long-wavelength cutoff of the wavenumber.

The analysis of their observations showed that the 17 and 34 GHz emissions exhibited synchronous quasi-periodical variations of the intensity at the loop top and footpoints of the flaring loop. Detailed comparison of the derived power spectra showed that the pulsations at the footpoints were almost synchronous with period of $P_2 = 8$–11 s. At the loop-top, the synchronism with the footpoint pulsations was not so prominent but nevertheless it existed on larger time scales of $P_1 = 14$–17 s. Reznikova and Melnikov pointed out that these properties of the pulsations indicate the possibility of simultaneous existence of two modes of sausage oscillations in the loop: the global mode with period of $P_1 = 14$–17 s and the nodes at the footpoints, and the harmonic with $P_2 = 8$–11 s.

3.2 Quasi-periodic Oscillations of the Slowly-varying Component

Using VLA data, Nindos (see also [50]) presented the first spatially resolved oscillations in total intensity and circular polarization of a stable

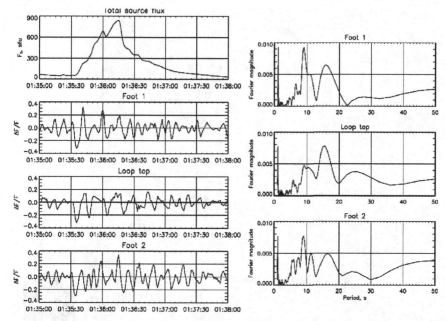

Fig. 2. NoRH observations of the 2000 January 12 flare at 17 GHz. The *left column* shows the time profiles and the *right column* shows the corresponding power spectra. The top left panel refers to the entire flaring loop while the other panels refer to the footpoints and loop top (from [49])

sunspot-associated gyroresonance source. Intermittent oscillations with, nevertheless, remarkable positional, amplitude, and phase stability were detected at 8.5 and 5 GHz. The spatial distribution of intensity variations was patchy and the location of the patches of strong oscillatory power was not the same at both frequencies. The strongest oscillations were associated with a small region where the 8.5 GHz emission came from the second harmonic of the gyrofrequency (see Fig. 3) while distinct peaks of weaker oscillatory power appeared close to the outer boundaries of the 8.5 and 5 GHz gyroresonance sources, where the emissions came from the third harmonic of the gyrofrequency. At both frequencies the oscillations had periods in the three–minute range: the power spectra showed two prominent peaks at 6.25–6.45 mHz and 4.49–5.47 mHz.

The authors checked the observed properties of the oscillations against model computations of the gyroresonance emission. They found that the oscillations are caused by variations of the location of the third and/or second harmonic surfaces with respect to the base of the TR, i.e. either the magnetic field strength or/and the height of the base of the TR oscillates. The best–fit model to the observed microwave oscillations can be derived from photospheric magnetic field strength oscillations with an rms amplitude of

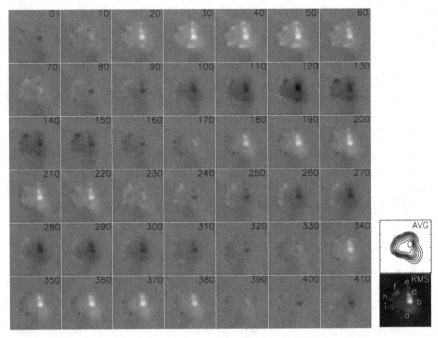

Fig. 3. 8.5 GHz I snapshot difference images obtained every 10 s between 22:19:00–22:26:00 UT. In the rightmost column the average and rms images derived from the entire 22:19:00–22:32:20 UT time series are presented. The difference images were produced after subtracting the average image from each snapshot image. *White* represents positive flux and *black* represents negative flux (from [50])

40 G or oscillations of the height of the base of the TR with an rms amplitude of 25 km. Furthermore small variations of the orientation of the magnetic field vector yielded radio oscillations consistent with the observed oscillations.

Gelfreikh reported quasi-periodic microwave oscillations of several types of stable sources. The oscillations were detected using a variety of instruments: the RATAN-600, NoRH, the Siberian Solar Radio Telescope (SSRT) and the Crimean RT-22. The most prominent oscillations were associated with sunspots showing periods of about 3 minutes. In some sunspots, however, oscillations with shorter and longer (up to 180 minutes) periods were detected. He detected oscillations not only above suspots but also above plages and pores. He pointed out the interesting result that sometimes within the same active region, different areas oscillate with different periods. Usually the plage-associated oscillations exhibited periods of about 10 minutes, but longer periods of about 50–80 minutes were not rare. He discussed possible mechanisms responsible for the observed oscillations and showed that acoustic modes with periods less than 1 min strongly dissipate in the lower solar corona due to thermal conduction losses while oscillations with periods of 10–40 s are associated with Alfvén disturbances.

4 Decimetric Pulsating Emission

4.1 Broad-band Pulsations

The decimetric pulsations are broadband emissions ($\Delta f/f$ about 0.5) with periodic or irregular short fluctuations. Sometimes they are quasi-periodic with pulses of 0.1 to 1 s separations, occurring in groups of some tens to hundreds and lasting a few seconds to minutes. As in the meter-wave range, the morphology shows considerable differences in modulation depths. Some decimetric events occur in continuum patches with drifting upper and lower frequency boundaries, with the drift being predominantly directed toward lower frequencies. Such patches are called drifting pulsating structures (DPS).

Traditionally, the emission of decimetric pulsations has been interpreted by a loss-cone instability of trapped electrons [6, 36]. An alternative interpretation of a decimetric drifting pulsating event (see Fig. 4) was presented by M. Karlický. This flare shows drifting pulsating structures of varying bandwidth in the 0.8–2 GHz range before the hard X-ray peak. In the main hard X-ray phase, the radio intensity increases and pulsations become less regular. *Yohkoh* soft X-ray images of this flare [53] revealed a nearly stationary low-lying flare loop and a plasmoid ejection, which was firstly seen at 9:24:40 UT at a height of $\approx 2 \times 10^4$ km. The ejected plasmoid was continuously accelerated up to a speed of $\approx$250 km s^{-1} at 09:25 and $\approx$500 km s^{-1} at 09:26 UT. The drifting pulsating structures of this event are interpreted in the framework of a model in which the radio pulsations are caused by quasi-periodic particle acceleration episodes that result from magnetic reconnection in a large-scale current sheet [34]. Under these circumstances, a possible reconnection model is the one that reconnection is dominated by repeated formation and subsequent coalescence of magnetic islands. This process is known as secondary tearing or impulsive bursty regime of reconnection. The continuously growing plasmoid is fed by newly coalescing islands. The plasmoid becomes strongly accelerated along the sheet and the pulsating particle source is located at the magnetic X line adjacent to the plasmoid. The radio source is then formed in or near the plasmoid. The frequency drift of the pulsating structures can naturally be attributed to the rise of the plasmoid in the corona, or to the growing upward expansion of the current sheet.

Other studies [27, 30, 31, 33] found drifting pulsating structures to be associated with the impulsive phase of flares, although they can occur before or after the HXR burst peak, and can be associated with plasma ejections seen in soft X-rays or EUV. At the workshop, Karlický presented one example of drifting pulsating structures observed during the X1.5 flare of March 18, 2003. The DPS was associated with a moving X-ray source mapped by RHESSI's 12–25 keV channel. Karlický presented a cartoon model which is shown in Fig. 5 attributing the different sources of DPS indicated by patch sequences observed in some decimetric dynamic spectra to the possible fragmentation of the evolving flare current sheet.

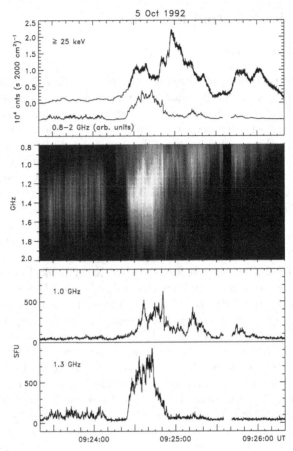

Fig. 4. Event of October 5, 1992 (from [34]). Dynamic radio spectrum (Ondřejov) (*middle*), two single-frequency cuts of the spectrum (*bottom*), and CGRO/BATSE/DISCSC hard X-ray light curve (*top*)

Khan showed the first imaging results of a decimetric drifting structure (see [33]). The DPS patch was imaged with the NRH at 327 MHz (see Fig. 6). It was accompanied by a soft X-ray plasmoid observed by *Yohkoh* SXT. Figure 6 indicates that the DPS radio source is located slightly above but overlapping with the plasmoid; however, the motion of the radio source was consistent with the plasmoid's motion.

The reported number of DPS events associated with plasmoid ejecta is small, and a meaningful statistical study is not yet available. Therefore, it is rather premature to argue that DPS events are commonly associated with plasmoid ejections (or vice versa). But if future studies confirm this association, DPS events could provide diagnostics of the flare current sheet evolution, especially the plasma density near the acceleration site and the temporal characteristics of particle acceleration by magnetic reconnection.

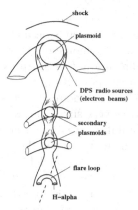

Fig. 5. Multiple decimetric pulsating source model as proposed by [30] giving an idea about a possible current sheet evolution in the main flare phase

4.2 Narrow-band Spikes

For decimetric spike bursts we refer to the review by [13]. Individual spikes are very short ($< 1\,$s), narrowband (some MHz only) bright emissions forming broadband clusters or patches of some tens to thousands during some tens of seconds to about a few minutes. Clusters are sometimes organized in small subgroups or chains. Sometimes they suggest quasi-periodic oscillatory temporal behavior at single-frequency records.

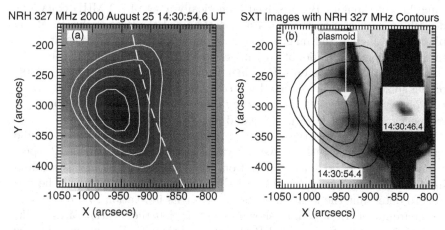

Fig. 6. Event of August 25, 2000 (from [33]). (**a**) NRH 327 MHz source during the drifting continuum emission (which, unfortunately, did not show clear pulsations at this frequency, see Fig. 1 of the original paper). (**b**) The contours of (**a**), now black, superposed on a composite SXT image. The radio source appears slightly ahead of the plasmoid

Using the spike's bandwidth, one may constrain its source size. If we assume that the emission frequency depends on a characteristic frequency (e.g. the local plasma frequency or a harmonic of the gyrofrequency) then an upper limit of spike's source dimension L is given by the product of the scale length of the characteristic frequency multiplied by $\Delta f/f$. Using typical values, [13] found that $L \leq 200$ km. Therefore, a successful model of narrowband spikes should be consistent with very small source sizes. Traditionally, models considered the loss-cone instability of trapped electrons producing electron cyclotron maser emission at the footpoints of flaring loops [44]. To avoid high magnetic fields in the source region, the model has been modified to emission of upper-hybrid and Bernstein modes [72]. This model can interpret the occasionally reported harmonic emission in decimetric spikes [14]. Alternatively, [67] proposed that the spikes come from sources located in the acceleration site of the flare and result from waves produced by the acceleration process (see also [37, 66]).

In the working group sessions, particular attention was paid to whether or not decimetric spikes are signatures of accelerated particles at the primary energy release site. This type of radio emission has been traditionally interpreted as a signature of highly fragmented energy release in flares [12]. Barta and Karlický discussed their model [10] in which dm-spikes are generated in the turbulent plasma of reconnection outflows. According to this model, superthermal particles are accelerated near the X-point of the magnetic field, or directly in the cascading MHD turbulence, and lead to kinetic instabilities. Karlický argued that the presence of narrowband dm-spikes in the dynamic spectrum of the X1.5 flare of March 18, 2003 should be considered as a further argument supporting the reconnection scenario for that flare. In the same line of thought, Barta solved numerically a set of 2D MHD equations describing magnetic reconnection and determined the time evolution of the plasma parameters and magnetic field. From these results, he calculated the radio emission due to double resonance instability in reconnection jets. He showed that, depending on MHD turbulence properties, either "lace bursts" or dm-spikes should be observed ("lace bursts" is a rare type of fine structure observed in the 1–2 GHz frequency range which is characterized by rapid frequency variations, both positive and negative; see [28]).

However, contradictory results were also reported. Khan discussed a radio spike event observed with the Astrophysical Institute of Potsdam (AIP) spectrometer. Using radio imaging data from the NRH and simultaneous soft and hard X-ray images from *Yohkoh* and EUV images from SOHO/EIT, he was able to determine the location of the spike bursts in relation to the flare and its environment. He found, as shown in Fig. 7, that the location of the spike bursts was remote from the HXR source. Additionally, Fig. 8 compares the (cluster-integrated) time profile of the radio flux of spike emission and the associated time profile of the hard X-ray emission. There is some correlation in the general behavior of the curves but not at all details. The movies presented by Khan indicated that the spike bursts occur at the site of compression of

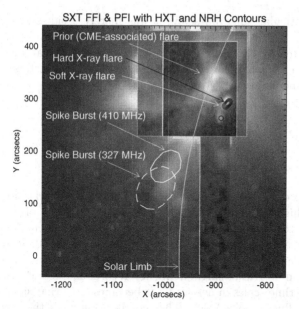

Fig. 7. Event of November 14, 1997. The *gray-scale* image is a composite of full-frame and partial frame *Yohkoh* SXT images. The *black contours* denote the hard X-ray image while *white contours* show the location of spike bursts according to Nançay maps (the *solid contour* corresponds to 410 MHz and the *dashed contour* to 327 MHz [32]

pre-existing loop structures. The compression was caused by a long-duration-event-associated CME which was launched nearby earlier (see also Fig. 7). The compression of the coronal loop structures can well result in magnetic reconnection leading to the spike burst event. Overall, the observations indicated

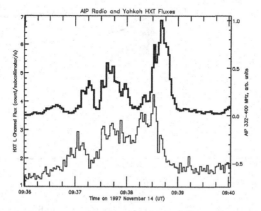

Fig. 8. Event of November 14, 1997. Frequency-integrated (332–400 MHz) radio spike flux taken from AIP spectral data (*thick line*), compared with *Yohkoh* HXR flux curve (14–23 keV) of the HXR source [32]

that that particular spike burst event was merely a side-show to the main energy release which was associated with the CME/LDE event. The observations reported by Khan do not contradict earlier observations by [15] who analyzed less convincing data suggesting that the spike clusters were located well outside the main energy release region of a flare.

5 Metric Pulsating Emission

5.1 Statistics

Magdalenič presented a statistical analysis of quasi-periodic oscillations selected from high time resolution single frequency records of Trieste Observatory (see also [40]). Figure 9 gives "frequency-of-occurrence" clusters over frequency for a 2.5-year sample (January 1998–June 2000). It is interesting that an $1/f$ dependence was found for the period range around 1 s. The high time resolution of Trieste single frequency records allowed Magdalenič to include shorter time scales in her study. It becomes clear that significant quasi-periodic flux fluctuations below 0.5 seconds occur over the whole spectral range, at least down to 200 MHz. If these very-short-period pulsations are due to MHD driver, then the Alfvén velocity can be used for a rough estimate of the typical source size (for the Trieste instrument's frequencies it lies between 500–1500 km s^{-1}; [70]). The resulting source dimensions corresponding

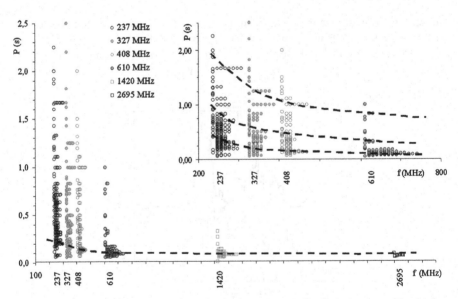

Fig. 9. Statistics of the dominant period of quasi-periodic fine structures in single-frequency Trieste Observatory polarimeter observations from January 1998 to July 2000 (courtesy of J. Magdalenič and P. Zlobec.)

to the very-short-period pulsations occurring at the low Trieste's frequencies are much smaller than the observed typical dimensions of coronal structures. Alternative mechanisms also need to be considered.

5.2 Broad Band Pulsations

In the working group sessions, metric broad band pulsations (BBP) with non-drifting spectral envelopes were discussed by Aurass. He presented a BBP event with a complex source configuration that occurred in a flaring sigmoidal loop system [8]. The analysis used radio spectral data from AIP while positional information was obtained from simultaneous NRH data. The pulsations' source sites were compared with soft X-ray images of the flare, and with force-free extrapolated coronal magnetic fields. Figure 10b shows an enlargement of the dynamic spectrum of the pulsations while in Fig. 10a we present the source configuration at three different NRH frequencies superposed on a soft X-ray image of the flaring active region.

The BBP sources occur in a diverging loop-like structure with a turning height of about 70 Mm. A surprising result of Aurass's analysis was that for a time interval of 90 s during the main flare phase, individual pulses consist of one source at lower frequencies, but are formed by two simultaneous

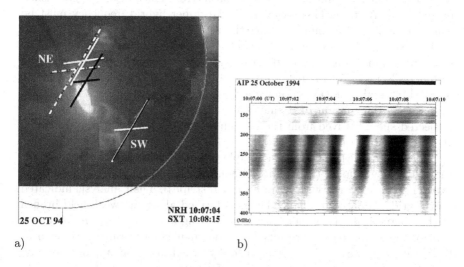

a)   b)

Fig. 10. Meter-wave pulsation detail from Fig. 1a. In (**a**): single-pulse source configuration at 327 MHz. NE: main source site. *Black cross*: onset, *white cross*: end of the pulse. *White-black stipple-dotted cross*: source sites at 236.6 and 164 MHz. SW: secondary pulsation source site with 90 s lifetime. The sizes of the crosses give the half widths of the sources. The distance between the NE and SW source sites is ≈ 0.12 R$_\odot$. In (**b**): the spectrum of this and neighboring pulses. Notice the J burst behavior at the low frequency edge

widely spaced sources at higher frequencies. These observations together with the event's magnetic field configuration suggest that the two high frequency sources must be fed by the same source of nonthermal electrons. This necessarily leads to the conclusion that in the given event, the broad band pulsations are due to a type III-like mechanism; the difference being in the source extent, and the density profile in the source. Furthermore, the beams exciting the pulse sequence are not accelerated within the radio-source-hosting configuration, but are injected near its strong-field footpoint. Note also that the double source of the pulsations was consistent with the overall evolution of the sigmoidal magnetic field configuration (see [8] for more details). It should be underlined here that in their pioneering work, [29] came to the same conclusion concerning the beam-driven nature of broad band pulsations and the external beam injection site.

5.3 Zebra Patterns

Originally zebra patterns were explained in terms of the excitation of Bernstein modes at harmonics of the gyrofrequency in a quasi-homogeneous compact source (e.g. [58]). Another approach assumes that their origin is associated with enhanced generation of plasma waves in regions of inhomogeneous coronal loops where the condition of double plasma resonance $f_{uh} = s f_B$ is satisfied (e.g. [35, 38, 73, 80]). This means the local upper hybrid frequency must be equal to the harmonic number times the local gyrofrequency.

Referring to the inherent relation between the zebra pattern formation and broad band pulsations, Aurass described in the working group discussions the results of [8] and [81]. Figure 11 gives the spectrum and positional data for a selected interval of the event shown in Fig. 1a. They found a good correlation between the inclination of a single zebra stripe around 164 MHz and the projected speed derived from the motion of a radio source appearing in the 164 MHz NRH positional data.

Zlotnik et al. [81] followed the double plasma resonance interpretation for the event they studied (according to [80], the Bernstein mode model cannot explain zebra patterns with more than 4–5 stripes). Their analysis indicated that a grid of double resonance layers is formed along the weak field branch of the magnetically asymmetric source-carrying loop structure. A frequency drift of the stripe pattern results from the changes of the magnetic field or temperature in the source volume. Field decrease leads to an apparent motion of the source of a given stripe to regions of higher electron density. The source motion's speed in the radio images is proportional to the frequency drift rate of the stripes. In the same manner acts the effect of plasma cooling.

Zlotnik et al. [81] tried to explain why sometimes (e.g. see our Fig. 1a) in dynamic spectra we observe broad band pulsations first and then zebra stripes at lower frequencies: they argued that this is is due to the higher threshold of the double resonance instability compared to the beam instability.

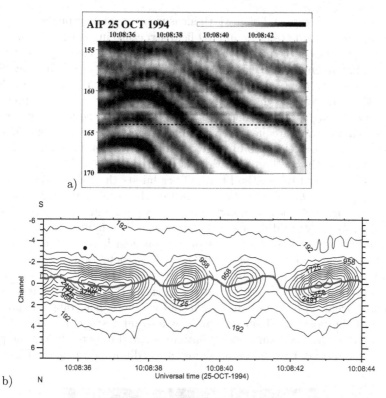

Fig. 11. Enlargement of the zebra pattern of Fig. 1a. In (**a**) spectrum (AIP). The *stippled line* denotes 164 MHz. In (**b**) the NRH 164 MHz contours of equal brightness (North-South array). The vertical axis is graded in instrument-related units, corresponding to a field of 1.28 $R_\odot$. The *fat line* gives the motion of the maximum of a Gaussian source model. Follow the stripes in the spectra and note the correlation of stripe inclination and source motion

5.4 Fiber Bursts

Fiber bursts are usually interpreted as the radio signature of whistler waves excited after their coalescence with Langmuir waves in loops with an unstable distribution of nonthermal electrons [35, 41, 42]. An alternative approach has been followed by [16] and [68] who invoked Alfvén solitons to explain fiber bursts. The two classes of models result in disturbances propagating either with the local whistler group velocity or with 1–3 times the local Alfvén velocity, respectively. Since at a given frequency, the modulator velocity is related to the burst's frequency drift rate, one can use the observed frequency drift rate to determine which model (if any) is more capable of explaining the observations.

Rausche presented observations of fiber bursts whose drift rates were consistent with the whistler wave model. He reported a promising new method that uses the drift rate and bandwidth measurements of fiber bursts as a probe of the magnetic field strength and 3D field structure in post-flare loops. The derived properties of the magnetic field can subsequently be compared directly with coronal magnetic field extrapolations. The method works when the fibers cross at least two frequencies for which positional information is available. It is based on the well-known property of whistler waves to propagate predominantly along magnetic field lines (e.g. [35]). Therefore, the radio source sites of fiber bursts at *two* frequencies determine (together with a coronal density model) a subset of field lines in the extrapolated coronal magnetic field. For more details on the method the reader is referred to the article by [9].

Rausche applied the method to the fiber bursts that occurred during the flare of April 7, 1997. Figure 12 summarizes the results. It shows the fiber burst-carrying field lines, forming an almost coplanar system during 1 hour of observations, overlaid on a soft X-ray image. The figure also shows their location with respect to the full set of coronal field lines obtained from the potential field extrapolation of a SOHO/MDI magnetogram. Figure 13 quantitatively displays the evolution of the fiber-burst carrying post-flare loop field lines over the 1 hour time interval. It is surprising that the fiber bursts intermittently occur within the same height and magnetic field range despite the growth of the post-flare loop system. It is also interesting that the whistler waves driving the fiber emission are initially excited near the top of the post-flare loops, and later in the event almost at medium loop heights. Figure 13

Fig. 12. *Yohkoh* postflare SXT image (07 April 1997, 16:40:28 UT, AlMg filter, 5.4 s exposure, N–upwards, W–to the right) with superposed potential field lines (*white*) and overplotted "mean fiber burst" field lines (*Yellow line*–10 min after impulsive phase, magenta–25 min after yellow, green–35 min after yellow). See also [9]

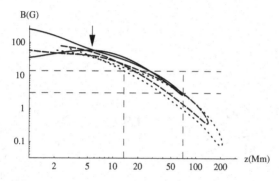

Fig. 13. Field strength versus height for the colored field lines in Fig. 12. The *solid, dashed and dotted curves* correspond to the *yellow, magenta, and green lines*, respectively of Fig. 12. All fiber bursts occur in the field strength–height range box formed by the *long-dashed lines*. The *arrow points* to a local field strength maximum occuring near the weak field end of the *continuous line* at height of about 5 Mm

further reveals the extremely low field strength near the top of fiber-burst carrying post-flare loops 1 hour after the impulsive flare phase. From the footpoint locations of the field lines selected with the fiber burst data, a post-flare loop footpoint expansion speed can be estimated which is consistent with EUV and X-ray data.

5.5 Source Model for Metric Fine Structure

It is well known (e.g. [62]) that broad band pulsations, fibers, and zebra patterns occur in complex radio bursts sometimes together (i.e. in the same flare event) and sometimes not. On the other hand, not all flares with complex radio signatures show well-developed fine structures. With Fig. 14 we outline the main properties of a source model for such fine structures following the results by [81].

What are the main features? All three types of fine structure occur in magnetically asymmetric loops (magnetic traps), at least at the beginning of fine structure emission during a given flare. The trap serves as the source of the accompanying broad band continuum emission by forming a loss-cone distribution of some background component of nonthermal electrons. Let us summarize some properties of the source model.

– A radio fine structure source needs nonthermal electrons. These seem to be injected during the associated flare mostly near the strong-field footpoint of the trap in the flaring active region. A possible source is the interaction between loops of different spatial scales as suggested e.g. by [26] and [51]. Zaitsev et al. [79] analyzed the efficiency of electron acceleration in such configuration.

– It is possible that the magnetic field near the strong-field footpoint interacts with part of a flaring structure (e.g. an approaching flare ribbon)

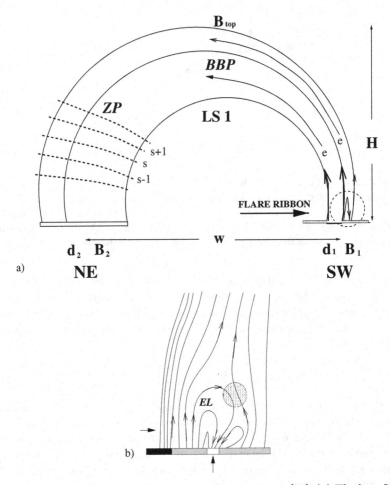

Fig. 14. Source model for metric continuum fine structures [81]. (**a**) The loop LS1 is the main source of BBP and ZP (*zebra patterns*): fast electron streams exciting BBP are injected at the SW footpoint. ZP stripes arise at the double plasma resonance levels (*stippled*) in the NE part of LS1. (**b**) Enlargement of the encircled SW region; the leading spot with polarity opposite to the polarity of the emerging loop (EL). The wavy circle is a site of reconnection and electron acceleration. This model refers to the analysis of the event shown in Fig.1a (see also [8])

and/or with emerging flux of parasitic polarity, possibly driving the loop interaction (with particle injection into the trap) into a periodic regime.
- Broad band pulsations seem to be due to repeated beam injection into the trap; this means metric pulsation pulses are type III-like emission within relatively small source structures. The radio source occurs between the strong-field footpoint and the top of the trap structure.

-   Unlike pulsations, fibers and zebra stripe sources are located at the weak-field branch of the trap. Altyntsev et al. [2] analyzed microwave zebra stripes and came to the same conclusion at much higher frequencies.
-   The analyzed events with meter-wave zebra stripes seem to support the models based on generation of electrostatic upper-hybrid waves at conditions of the double plasma resonance (e.g. [80, 81]).
-   For the analyzed events with decimeter/meter-wave fiber bursts, the Alfvén soliton fiber burst model of [68] can be ruled out. Fibers seem to be associated with whistler wave packets in coronal loops as proposed by [35] and modified by Mann et al. [41, 42].

6 Summary and Future Work

In this section we present questions rather than conclusions. A firm conclusion, however, is that radio pulsations provide important diagnostics of the physical conditions in the corona. Most observations of radio pulsations come from non-imaging spectral data. The last few years, however, several studies that combine spectral data with simultaneous images obtained from the NoRH (microwaves) or NRH (decimetric-metric wavelengths) have appeared resulting in an accumulation of several well-observed events. These observations do not suffer from the positional ambiguities of earlier studies and have yielded significant progress in the subject. Furthermore, the combination of radio data with soft X-ray and EUV images from space telescopes offers the opportunity for a more complete view of the coronal configuration and the processes involved.

In comparing observations with theory, one has the feeling that the coupling between them has not reached the desired level yet. The observers often use formulas giving the periods of some MHD oscillation modes and compare them with their data. Of course, this is useful because it may provide diagnostics about the coronal plasma and magnetic field; however it does not tell much about the physical processes leading to radio emission modulation. On the other hand, theoretical studies need to take into account the latest observational results.

Clearly, we need to learn more about fundamental processes leading to flares and coronal mass ejections. For example, can we use the "standard reconnection model" for the interpretation of dm drifting pulsating structures and spikes? In the working group sessions, such attempts were made, but only a couple of events were considered. In any case we cannot exclude other approaches. For instance, the appearance of decimetric spikes as evidence for the existence of multiple current sheets formed stochastically, versus evidence for spikes as a by-product of deterministic large-scale dynamic processes, needs to be tested. Furthermore, in the working group sessions, evidence for the beam-driven nature of broad band pulsations was presented for one event.

A similar in-depth study needs to be done for several other BBP events before reaching firm conclusions.

In the working group sessions, several participants demonstrated how to exploit the information provided by radio pulsations in order to derive information about the coronal magnetic field. However, this process is not trivial and requires modeling at some stage. For example, the quasi-periodic fluctuations of strong steady sunspot-associated gyroresonance sources combined with models of the gyroresonance emission may yield accurate estimates about the underlying magnetic oscillations. Also, the combination of fiber burst data with magnetic field extrapolations provides a powerful tool to probe the 3D structure of post-flare loops.

Regarding microwave pulsations, an effort should be made to integrate the observations with longer wavelength radio, and coronal observations from space instruments. Observations of spatially resolved microwave oscillations (both from flares and the slowly-varying component) were presented in the meeting. This is a significant improvement but we believe that the use of simultaneous soft X-ray or/and EUV observations will provide additional progress. We do not know yet much about the relation between flare-related microwave oscillations and pulsations at metric wavelengths. In the study of decimetric and metric pulsations we need to clarify the associated magnetic field configuration. For example, in the working group sessions, there was a debate whether dm spike bursts are signatures of accelerated electrons at the primary energy release site or just a by-product of the main flare. Also, the possible association of dm drifting pulsating structures with plasmoid ejections needs to be tested with more observations. Here, the lack of simultaneous spectroscopic and imaging observations in the decimetric range is the main obstacle.

With the future development of the "Frequency Agile Solar Radiotelescope", imaging spectroscopy over a wide spectral range in radio will be achieved. This combined with the forthcoming space missions promises exciting new developments on the subject. Independently, broadband spectroscopy, possibly combined with extremely enhanced spectral and time resolution in subbands, will also remain a useful research tool in the future.

Acknowledgements

AN would like to thank Prof. C.E. Alissandrakis for useful discussions. He also acknowledges support from European Union's grant INTAS 00–543. HA is grateful to M. Karlický for discussions about dm pulsations. He also acknowledges the European Office for Aerospace Research and Development for its support in maintaining the solar radio spectral observations at Potsdam.

References

1. A. Abrami: Solar Phys. **11**, 104 (1970)
2. A.T. Altyntsev, A.A. Kuznetsov, N.S. Meshalkina, S. Natalya, Y. Yan: Adv. Space Res. **35**, 1789 (2005)
3. A. Asai, M. Shimojo, H. Isobe, and 4 coauthors: ApJ **562**, L103 (2001)
4. M. Aschwanden: Solar Phys. **111**, 113 (1987)
5. M.J. Aschwanden, L. Fletcher, C.J. Schrijver, D. Alexander: ApJ **520**, 880 (1999)
6. M. Aschwanden, A.O. Benz: ApJ **332**, 447 (1988)
7. M. Aschwanden, V.M. Nakariakov, V.F. Melnikov: ApJ **600**, 458 (2004)
8. H. Aurass, K.-L. Klein, E. Ya. Zlotnik, V.V. Zaitsev: A&A **410**, 1001 (2003)
9. H. Aurass, G. Rausche, G. Mann, A. Hofmann: A&A **435**, 1137 (2005)
10. M. Barta, M. Karlický: A&A **379**, 1045 (2001)
11. L.R. Bellot-Rubio, M. Collados, B. Ruiz Cobo, I. Rodriguez Hidalgo: ApJ **534**, 989 (2000)
12. A.O. Benz: Solar Phys. **96**, 357 (1985)
13. A.O. Benz: Solar Phys. **104**, 99 (1986)
14. A.O. Benz, M. Güdel: Solar Phys. **111**, 175 (1987)
15. A.O. Benz, P. Saint-Hilaire, N. Vilmer: A&A **383**, 678 (2002)
16. T.E.X. Bernold, R. Treumann: ApJ **264**, 677 (1983)
17. N. Brynildsen, T. Leifsen, O. Kjeldseth-Moe, P. Maltby, K. Wilhelm: ApJ **511**, L121 (1999a)
18. N. Brynildsen, O. Kjeldseth-Moe, P. Maltby, K. Wilhelm: ApJ **517**, L159 (1999b)
19. F. Dröge: A&A **57**, 285 (1977)
20. Ø. Elgarøy: *Solar Noise Storms* (Pergamon Oxford 1977), pp. 177–185
21. G.D. Fleishman, Q.J. Fu, G.L. Huang, V.F. Melnikov, M. Wang: A&A **385**, 671 (2002)
22. A. Fludra: A&A **344**, L75 (1999)
23. G.B. Gelfreikh, V. Grechnev, T. Kosugi, K. Shibasaki: Solar Phys. **185**, 177 (1999)
24. N. Gopalswamy, E.J. Schmahl, M.R. Kundu: BAAS **25**, 1396 (1993)
25. V.V Grechnev, S.M. White, M.R. Kundu: ApJ **588**, 1163 (2003)
26. Y. Hanaoka: Double-Loop Configuration and Its Related Activities. In: *ASP Conf. Series* Vol. 111, ed. by R.D. Bentley, J.T. Mariska (Astron. Soc. of the Pacific, San Francisco 1996) pp. 200–205
27. K. Hori: Study of Solar Decimetric Bursts with a Pair of Cutoff Frequencies. In: *Solar Physics with Radio Observations* NRO Report **479**, ed. by T. Bastian, N. Gopalswamy, K. Shibasaki (1999), pp. 267–271
28. K. Jiřička, M. Karlický, H. Mészárosová, V. Snižek: A&A, **375**, 243 (2001)
29. K. Kai, A. Takayanagi: Solar Phys. **29**, 461 (1973)
30. M. Karlický: A&A **417**, 325 (2004)
31. M. Karlický, F. Fárník, and H. Mészárosová: A&A **395**, 677 (2002)
32. J.I. Khan, H. Aurass, H.: Observations of the Coronal Dynamics Associated with Solar Radio Spike Burst Emission, in preparation (2005)
33. J.I. Khan, N. Vilmer, P. Saint-Hilaire, A.O. Benz: A&A **388**, 363 (2002)
34. B. Kliem, M. Karlický, and A.O. Benz: A&A **360**, 715 (2000)
35. J. Kuijpers: Solar Phys. **44**, 173 (1975)

36. J. Kuijpers: Theory of Type IVdm Bursts. In: *Radiophysics of the Sun*, IAU Symp. **86** ed. by M. Kundu and T.E. Gergely (Reidel, Dordrecht 1980), pp. 341–361

37. J. Kuijpers, P. van der Post, C. Slottje: A&A **103**, 331 (1981)

38. V.G. Ledenev, Y. Yan, Q. Fu: Chin. Journ. of Astron. Astrophys **1**, 475 (2001)

39. B.W. Lites: Sunspot oscillations—Observations and implications. In: *Sunspots: Theory and observations; Proceedings of the NATO Advanced Research Workshop on the Theory of Sunspots* (Cambridge, United Kingdom, A93-47383 1992), pp. 261–302

40. J. Magdalenić, P. Zlobec, M. Messerotti, B. Vrsnak: ESA SP-**506**, 335 (2002)

41. G. Mann, M. Karlický, U. Motschmann: Solar Phys. **110**, 381 (1987)

42. G. Mann, K. Baumgärtel, G.P. Chernov, M. Karlický: Solar Phys. **120**, 383 (1989)

43. D.J. McLean, K.V. Sheridan: Solar Phys. **32**, 485 (1973)

44. D.B. Melrose, G.A. Dulk: ApJ **259**, 844 (1982)

45. H. Nakajima: Solar Phys. **86**, 427 (1983)

46. V.M. Nakariakov, L. Ofman, E.E. Deluca, B. Roberts, J.M. Davila: Science **285**, Iss. 5429, 862 (1999)

47. V.M. Nakariakov, L. Ofman: A&A **372**, 53 (2001)

48. V.M. Nakariakov, A.V. Stepanov: this volume

49. V.M. Nakariakov, V.F. Melnikov, V.E. Reznikova: A&A **412**, L7 (2003)

50. A. Nindos, C.E. Alissandrakis, G.B. Gelfreikh, V.M. Bogod, C. Gontikakis: A&A **386**, 658 (2002)

51. M. Nishio, K. Yaji, T. Kosugi, H. Nakajima, T. Sakurai: ApJ, **489**, 976 (1997)

52. A. Norton, R.K. Ulrich, R.I. Bush, T.D. Tarbell: A&A **518**, L123 (1999)

53. M. Ohyama, and K. Shibata: ApJ **499**, 934 (1998)

54. K. Yu. Platonov, G.D. Fleishman: Physics–Uspekhi **45** (3), 235 (2002)

55. Z. Qin, C. Li, Q. Fu, Z. Gao: Solar Phys. **163**, 383 (1996)

56. B. Roberts, P.M. Edwin, A.O. Benz: Nature, **305**, 688 (1983)

57. B. Roberts, P.M. Edwin, A.O. Benz: ApJ **279**, 857 (1984)

58. H. Rosenberg: Sol. Phys., **25**, 188 (1972)

59. I. Rüedi, S.K. Solanki, J.O. Stenflo, T. Tarbell, P.H. Scherrer: A&A **335**, L97 (1998)

60. J.I. Sakai, C. de Jager: Space Sci. Rev. **77**, 1 (1996)

61. K. Shibasaki: ApJ **550**, 326 (2001)

62. C. Slottje: *Atlas of Fine Structures of Dynamic Spectra of Solar Type IV-dm and Some Type II Radio Bursts* (Utrecht, Publ. Utrecht University 1981)

63. J. Staude: Sunspot Oscillations. In: *3rd Advances in Solar Physics Euroconference: Magnetic Fields and Oscillations*, ed. by B. Schmieder, A. Hofmann, J. Staude, (ASP Conf. Series 184, 1999), pp. 113–123

64. A.V. Stepanov: Res. Geomagn., Aeronomy, and Solar Phys. (in Russ.) **54**, 141 (1980)

65. A.V. Stepanov, S. Urpo, V.V. Zaitsev: Solar Phys. **140**, 139 (1992)

66. A.V. Stepanov, B. Kliem, A. Krüger, J. Hildebrandt, V.I. Garaimov: ApJ **524**, 961 (1999)

67. T. Tajima, A.O. Benz, M. Thaker, J.N. Leboeuf: A&A **353**, 666 (1990)

68. R.A. Treumann, M. Güdel, A.O. Benz: A&A **236**, 242 (1990)

69. G. Trottet, A. Kerdraon, A.O. Benz, R. Treumann: A&A **93**, 129 (1981)

70. B. Vršnak, J. Magdalenić, H. Aurass, G. Mann: A&A, **396**, 673 (2002)

71. S.M. White, M.R. Kundu: Solar Phys. **174**, 31 (1997)
72. A.J. Willes, P.A. Robinson: ApJ **467**, 465 (1996)
73. R.M. Winglee, G.A. Dulk: ApJ **307**, 808 (1986)
74. C.W. Young, C.L. Spencer, G.E. Moreton: ApJ **133**, 243 (1961)
75. V.V Zaitsev: Solar Phys. **20**, 95 (1971)
76. V.V. Zaitsev, A.V. Stepanov: Soviet Astron. Letters **8**, 132 (1982)
77. V.V. Zaitsev, A.V. Stepanov: Soviet Astron. Letters **15**, 154 (1989)
78. V.V. Zaitsev, A.V. Stepanov, S. Urpo, S. Pohjolainen: A&A **337**, 887 (1998)
79. V.V. Zaitsev, E.Ya. Zlotnik, H. Aurass: *Astronomy Letters* **31**, 285 (2005)
80. V.V. Zheleznyakov, E.Ya. Zlotnik: Solar Phys. **43**, 431 (1975)
81. E. Ya. Zlotnik, V.V. Zaitsev, H. Aurass, G. Mann, A. Hofmann: Astron. Astrophys. **410**, 1011 (2003)

Index

accelerated electrons, 124
  anisotropy, 78
  bump-on-tail distributions, 167
  distribution functions, 167
  energy content, 35, 44, 58
  energy spectrum, 35, 60, 96–97
    and photon spectrum, 74–77
    cutoff, 73
    dip, 77
    interacting vs. escaping, 78–79
  loss-cone distributions, 167, 172
  power-law energy distribution, 176
  ultrarelativistic, 90
accelerated ions, 51
  energy spectrum, 97
accelerated particles
  distribution function, 95
  energy content, 16
  energy distribution, 20
  number, 17
  precipitation, 26
  trapping, 26, 29, 96
AD Leo, 246
AIP Potsdam spectrometer (OSRA),
    89, 254, 264, 265, 267, 269
Alfvén solitons, 269, 273
Alfvén speed, 89, 108, 110, 128, 130,
    146, 149, 157, 158, 174, 191,
    212–214, 224, 226, 237, 247, 255,
    258, 266, 269
Alfvén-Carlquist Model, 242
Alfvén-Carlquist model, 244
anomalous resistivity, 244

Atacama Large Millimeter Array
    (ALMA), 82, 98
avalanche, 24

ballooning instability, 244
ballooning mode, 233–234, 238–242
beam instability, 268
betatron acceleration, 209
bi-Maxwellian distribution, 191, 195,
    196, 201
Buneman instability, 244

Chinese Radio Heliograph, 94
chromospheric evaporation, 43, 71, 73,
    86, 255
collisions, 26, 191, 192, 198–199, 201
  mean free path, 192
Compton GRO/BATSE, 93
Compton scattering, 77
conductive evaporation, 71
conductivity, 234, 243
coronal dimming, 116, 125, 126
coronal hole, 193, 194
coronal magnetic field, 20, 97, 132,
    269–271
  extrapolation, 23, 27, 267, 270, 274
  in post flare loops, 270
coronal mass ejection (CME), 16, 89,
    94, 108, 116, 204–207
  acceleration, 85, 128, 129, 144, 145,
    147, 206, 208, 209
  and wave phenomena, 124–127
  breakout model, 205
  energy, 59

interaction, 148–152
onset, 90
speed, 144, 145, 147, 207, 208
tether cutting model, 205
width, 144, 145, 147, 207, 208
coronal seismology, 131–132, 221, 229,
    233, 235, 247, 252
CORONAS/RESIK, 36
Coulomb collisions, 55, 192, 198
Cowling resistance, 242
cross-field diffusion, 27
cross-field drift, 163, 166, 187
CSHKP model, 85
current sheet, 1, 16–24, 27, 29, 30,
    40–43, 61, 70, 74, 87–88, 95, 186,
    255, 261–263, 273
  free energy, 23
  stability, 17

double plasma resonance, 264, 268, 272,
    273

electric circuit, 233, 242–247
electric current, 242–247
electric field, 20, 200, 243
  ambipolar, 200
  motional, 186, 187
  stochastic, 19
electron beam, 97, 165, 204
  density, 70
  filamentation, 74
  injection area, 74
  rate, 70, 74
electron cyclotron maser, 264
electron distribution function
  gyrotropic, 195
  halo, 192, 195, 196
  solar wind, 192–196
  strahl, 192, 195
  thermal core, 192, 195, 196
electron foreshock, 165, 172, 173
energy losses
  collisional, 68, 75, 77
  non collisional, 69
EUV irradiation, 124
extended flare, 88–90

fast solar wind, 191, 193

Fermi acceleration, 161, 173, 175, 177,
    182, 184
filament, 122, 131
flare ejecta, 94, 127
flare kernels
  motion, 206, 210
flare ribbons, 50, 69, 70, 84–87, 95, 271
flare spray, 121, 127, 129
flare/CME
  breakup model, 17–20, 27, 28
  loop model, 19–22, 28, 29
flare/CME energetics, 58–60
flare/CME event
  1991 March 24, 245
  1992 October 05, 93, 262
  1994 October 25, 254, 267, 269
  1996 August 19, 210
  1997 April 07, 114, 270, 271
  1997 November 03, 116, 120
  1997 November 04, 122
  1997 November 14, 265
  1997 October 07, 210
  1998 August 08, 119
  1998 August 24, 118
  1998 May 02, 113, 115, 122
  1998 May 08, 238
  2000 August 25, 263
  2000 December 18, 211
  2000 February 08, 205
  2000 January 12, 236, 259
  2000 March 13, 96
  2001 April 12, 91
  2001 April 14, 149
  2001 April 15, 149
  2001 May 07, 142
  2002 April 14, 89
  2002 April 14/15, 70, 73
  2002 April 15, 41, 42, 87
  2002 April 16, 43
  2002 April 21, 59, 94
  2002 February 20, 43–45, 59, 67, 68
  2002 February 26, 45, 58
  2002 July 17, 86
  2002 July 18, 210
  2002 July 20, 150
  2002 July 23, 38, 46, 47, 49–51, 53–55,
      57, 59, 66, 68, 69, 77, 94, 97, 151
  2002 March 14, 69
  2003 April 26, 36, 37

2003 March 18, 261, 264
2003 November 03, 88–90, 93, 117, 209
2003 November 04, 92
2003 November 05, 60
2003 October 28, 52–54, 57, 91, 92
2004 July 05, 39, 40
flux rope, 125, 205, 206
Fokker-Planck equation, 22, 198
fragmentation of energy release, 16, 26, 27, 29
Frequency Agile Solar Radio Telescope (FASR), 26, 94, 98, 274

gamma rays, 26, 51–58
  0.511 MeV line, 51–53, 55, 91
  2.223 MeV line, 52, 54–56
    imaging, 57, 66
  bremsstrahlung, 52, 55
  de-excitation lines, 53–54
  from pion decay, 91
  imaging, 56–58, 60
  line broadening, 54
  nuclear lines, 33, 55
  redshift, 54, 55, 60
geocentric solar ecliptic (GSE) system, 181
Giant Metrewave Radio Telescope (GMRT), 82

Hall current, 243
hard X-rays, 26, 43–51, 83–94, 261, 262, 264, 265
  albedo, 46, 77, 78
  and microwaves, 27, 95–97
  bremsstrahlung, 33, 57
  corona, 41, 42, 50, 71
  directivity, 78
  partially ionised plasma, 46, 69
  photon spectrum, 26, 45–47, 52, 60, 74
  plasma density, 67, 70
  pulsations, 233
  spectral analysis, 45
  spectral analysis techniques, 76
  thermal emission, 35, 43, 44, 48, 84, 87, 88
  thermal/nonthermal, 47–48, 60

thick target model, 45, 65, 67, 68, 70, 77, 88, 95, 96
heat flux, 199
hot flow anomalies, 184–188

impulsive flare, 35, 41, 47, 48, 50, 88, 94, 96
interplanetary magnetic field, 162
ion acceleration
  diffuse ions, 176, 180–183
  power law energy spectrum, 177, 178
ion foreshock, 181
ion-neutral collisions, 244
ISEE 3, 75

jet, 17, 58, 127, 209
Joule dissipation, 244

kappa function, 195–201
kink instability, 205
kink mode, 131, 222, 224, 231, 248, 255
Knudsen number, 192–194, 198, 201
Köln Observatory for Submillimeter and Millimeter Astronomy (KOSMA), 90, 91

Liouville's theorem, 168, 170, 172, 197, 200
loop, 20, 42, 87, 140, 151, 154, 205, 206, 209–211
  above looptop X-ray source, 61, 65, 88
  curvature, 223, 233
  erupting, 127
  hard X-rays, 48, 70
  inductance, 244
  microwave footpoint, 96
  microwave looptop, 26, 96
  oscillations, 130–131
  oscillations (EUV), 231
  post flare loop, 270, 274
  transequatorial, 125, 205
  UV footpoints, 86
  waves, 222
  X-ray footpoints, 26, 41, 43, 44, 46–51, 65, 67–70, 84–90, 95, 209
    motion, 49, 50, 69, 70, 86, 87
  X-ray looptop, 41, 46, 65, 68–70, 209
loop-loop interaction, 255, 256, 271, 272

loss cone distribution, 271
loss cone instability, 261, 264
loss of equilibrium, 94, 205

Mach number, 111, 158, 164
    Alfvénic, 120, 149, 162
    magnetosonic, 121
magnetic mirroring, 55, 96, 167–172
magnetic moment, 168
magnetic reconnection, 1, 23, 49, 50, 61,
        65, 69, 70, 85–88, 94, 95, 108, 140,
        206–207, 255, 261, 262, 265, 272,
        273
    guide field, 95
    jet, 264
    outflow, 209, 264
    rate, 51, 206, 210
magnetic shear, 20
magnetic stress, 24, 28
magnetosonic speed, 110, 121, 122, 132,
        212–214
Maxwellian distribution, 192, 195–197
Metsähovi radio telescope, 245
MHD turbulence, 26, 264
microcalorimeter arrays, 83
microflare, 39–40, 71, 72, 83–84
    heating, 71
    mass supply, 71

Nançay Radio Heliograph (NRH), 83,
        88, 253, 262, 264, 267, 268, 273
nanoflare, 71, 83
Neupert effect, 72–74
    plasma cooling, 73
Nobeyama Radio Heliograph (NoRH),
        81, 83, 96, 98, 235, 247, 256,
        258–260, 273

Ohmic dissipation, 242
Ondřejov radio spectrograph, 93, 262
Owens Valley Radio Observatory
        (OVRO), 83

Pannekoek-Rosseland potential, 197
particle acceleration, 16, 22–26, 46, 58,
        65, 82, 88, 90, 94–98, 130, 209–211
    diffusive, 20
    MHD turbulence, 20
    rate, 17

stochastic, 20, 58, 67
photoelectric absorption, 77
pitch angle scattering, 55, 173, 176
plasma beta, 110, 164
plasma heating, 26
plasmoid, 41, 94, 127, 211, 255, 261–263,
        274
positronium, 52, 53
proton beam, 191

radar experiments, 214
radio emission, 26–29, 88–94, 162, 164,
        165, 172
    broadband pulsations, 261–262,
        267–268, 271
    by positrons, 91
    continuum, 89
    drifting pulsating structure (DPS),
        93, 94, 261, 273, 274
    fiber burst, 253, 268–271, 273, 274
    free-free, 91
    gyroresonance, 256–260, 274
    gyrosynchrotron, 89–91, 96–98, 229,
        257
    intermediate drift burst (=fiber
        burst), 253
    lace burst, 264
    microwave pulsations (active regions),
        258–260
    microwave pulsations (flares), 229,
        233, 235–236, 245, 257–258
    microwaves, 26–27
    noise storm, 29, 82, 253
    plasma diagnostics, 211–214
    pre-flare, 28–29
    pulsations, 233, 247
    quasiperiodic pulsations, 222, 247
    reverse slope, 2
    spikes, 27, 253, 262–266, 273, 274
    synchrotron, 91, 93, 96
    thermal, 91
    transition radiation, 258
    type I, 29
    type II, 5–6, 27–28, 89, 94, 107, 108,
        119–120, 122, 124–129, 131–133,
        140–146, 162, 172, 210, 212, 213,
        253, 254
    and CMEs, 128, 142–146, 208–209
    and flares, 128, 157

starting frequency, 141
type III, 2, 27, 89, 97, 130, 204, 211, 214, 253, 268, 272
stochastic model, 27
type IV, 222, 253, 254
type J, 205
zebra pattern, 253, 254, 268, 271–273
Razin suppression, 98
resistivity, 21, 26
return current, 74
RHESSI, 34–39, 41, 42, 44, 45, 49, 50, 52, 55, 66–70, 72, 75, 83, 84, 87–91, 94, 95
RLC oscillations, 244–247
RT-22 radio telescope, 260

sausage mode, 224, 226–228, 236–238, 255, 258
self-organised criticality, 23–26
shock drift acceleration, 210
shock wave, 17, 20, 22, 109–112, 121–124, 141, 161–188, 207–211, 213
  at reconnection outflow, 95, 210
  blast wave, 108, 113, 127, 128, 131, 145, 157, 208
  bow shock, 112, 161–188, 208
  Earth, 162
  CME driven, 108, 127–130, 145, 156–157, 188, 208
  collisionless, 163
  compression ratio, 177
  cross-shock potential, 168–170
  curvature, 171, 175
  de Hoffman-Teller frame (HTF), 168
  diffusive particle acceleration, 177
  diffusive transport, 178, 182–184
  driven, 111, 128, 214
  electron acceleration, 164–176
  fast mode, 122, 140, 210
  flare related, 108, 127–130, 157, 188, 209, 214
  foot, 163, 174
  foreshock, 163–165, 167
  ion acceleration, 176–184
  ion reflection, 163
  non-stationary, 172
  normal incidence frame (NIF), 168
  overshoot, 163, 167, 174

particle acceleration, 85, 94, 130, 140, 161–188, 210, 211
  seed particles, 150
particle injection, 165, 178, 179, 182, 187, 188
piston, 111, 112, 125, 127, 208
pulsations, 178–182
quasi-parallel, 162, 163, 176–184
quasi-perpendicular, 162–176, 186–188
ramp, 163, 167, 174
stand-off distance, 162
thermalization, 163, 178, 179, 181
Siberian Solar Radio Telescope (SSRT), 83, 260
SLAMS (short, large amplitude magnetic structure), 179
soft X-rays
  bremsstrahlung, 33
  Fe/Ni lines, 36, 48
SoHO/LASCO coronograph, 89
solar energetic particles (SEP), 130, 140, 211
  and type II bursts, 146–148
  electrons, 48, 78–79, 89
  gradual, 140
  impulsive, 140
  intensity, 147, 152–154
    and active region area, 154
    and CME interaction, 148–154, 158
    and flare size, 149
  interplanetary propagation, 79
Solar Submillimeter Telescope (SST), 83, 90, 92
solar wind
  exospheric model, 192, 199–201
  fluid model, 191, 196
solar wind acceleration, 193
sound speed, 226, 231, 233
Spitzer conductivity, 244
Spitzer-Härm theory, 193, 199
stellar flare, 247
submillimetre emission, 90–93
  modulation, 93
  spectral upturn, 91
sunspot, 256, 257, 259, 260, 274

tangential discontinuity (TD), 186–188
thermal conduction, 124

thermal plasma, 35–39
tokamak, 234
TRACE, 41, 49, 66, 69, 86
transition region, 256
Trieste Observatory polarimeter, 254,
     266, 267
two-ribbon flare, 206

velocity filtration, 196–198
Very Large Array (VLA), 81, 82, 257,
     258
viscosity, 132, 235

waves
  acoustic wave, 231, 256, 257, 260
  Alfvén wave, 20, 110, 225, 229, 255,
     260
  Bernstein modes, 264, 268
  coronal waves, 112–134
  EIT waves, 108, 113–128, 130, 132,
     133, 209, 211, 222
    brow waves, 114
    S-waves, 114
  fast mode, 110, 121, 224, 225,
     227–229, 255
  Helium I, 117–118, 123
  Langmuir wave, 97, 165, 170
  leaky mode, 227
  longitudinal wave, 229–233, 248

lower hybrid, 173
magnetic waves, 177, 185
MHD waves, 21, 109–112, 120–124,
     126, 221–222, 251, 252, 255, 257,
     273
  damping, 234–236
  of a plasma cylinder, 223–229
Moreton waves, 108, 112–130, 132,
     133, 208
radial mode, 227, 247
radio band, 118–119, 123
simple waves, 111, 121, 208
slow mode, 110, 224, 231, 255
soft X-rays, 116–117, 123
surface wave, 174
trapped mode, 224, 226, 227
ULF waves, 176, 178, 179, 181
upper hybrid, 257, 264, 273
whistler, 173, 174, 269, 270, 273
white light flare, 47, 92
Wigner-Ville transform, 245
Wind/WAVES experiment, 89

X-rays, 233

Yohkoh/SXT, 93

Zürich radio spectrograph, 89